Math for All Learners Geometry

by
Pam Meader and Judy Storer

illustrated by
Jennifer DeCristoforo

J. WESTON WALCH PUBLISHER
Portland, Maine

User's Guide
to
Walch Reproducible Books

As part of our general effort to provide educational materials that are as practical and economical as possible, we have designated this publication a "reproducible book." The designation means that purchase of the book includes purchase of the right to limited reproduction of all pages on which this symbol appears:

Here is the basic Walch policy: We grant to individual purchasers of this book the right to make sufficient copies of reproducible pages for use by all students of a single teacher. This permission is limited to a single teacher and does not apply to entire schools or school systems, so institutions purchasing the book should pass the permission on to a single teacher. Copying of the book or its parts for resale is prohibited.

Any questions regarding this policy or requests to purchase further reproduction rights should be addressed to:

Permissions Editor
J. Weston Walch, Publisher
321 Valley Street • P. O. Box 658
Portland, Maine 04104-0658

1 2 3 4 5 6 7 8 9 10

ISBN 0-8251-4260-1

P. O. Box 658 • Portland, Maine 04104-0658
www.walch.com

Printed in the United States of America

Contents

To the Teacher

Since the early 1970s, we have been teaching math to learners of all ages, from young children to adults, who represent many different cultures and socioeconomic backgrounds. We believe that all learners *can* do math by first overcoming any math anxiety and then by participating in meaningful, cooperative learning activities that relate to various learning modalities (e.g., auditory, visual, kinesthetic, and tactile) of each learner.

With the release of the NCTM Standards in 1988 and in 2000 the NCTM Principles and Standards for School Mathematics, teaching mathematics has become more student-driven and hands-on. Emphasis on deductive and inductive reasoning through a discovery process enables a student to truly *understand* mathematics. We feel the labs presented in this book address these concerns. First, our labs address students' various learning styles. We provide hands-on activities of measuring, constructions, etc., which address the kinesthetic learner. We give opportunities to write and communicate ideas and to visualize concepts, thus including the visual learner. Finally, we furnish opportunities for group discussions and talking through problems, enabling the auditory student to be involved.

The study of Euclidean geometry lends itself to discovery of theorems through hands-on applications. Students who derive their own meaning for various theorems will own them and will understand what the theorems mean.

We hope you enjoy trying these activities with your students. We believe learning should be learner-centered, not teacher-driven. As quoted in the NCTM Principles and Standards, "students should conduct . . . explorations which will allow them to develop a deeper understanding of important geometric ideas. . . ."

—Pam and Judy

Ratio

Activity 1

Teacher Page

The Golden Ratio

Learning Outcomes

Students will be able to

- understand what the golden ratio is.
- understand the concept of constant ratio.

Overview

Students will measure rectangular objects and express the relation of length to width as a ratio.

Time Requirements

30–45 minutes

Group Size

Pairs

Materials

- ruler
- meterstick or yardstick
- 3 × 5 index cards
- 5 × 8 index cards
- playing cards
- notebook paper
- picture frames and envelopes of various sizes
- lab sheet
- calculator

Procedure

1. Before beginning the lab, a discussion of what a constant ratio is and how to change ratios into decimals is needed. Also the teacher may wish to give some background on the golden ratio. The Parthenon has many rectangles in its structure that demonstrate the golden ratio.
2. Pass out lab sheets, calculator, small index card, large index card, a playing card, a picture frame, notebook paper, and an envelope. Have students measure their objects and post on the lab sheet. Students will also calculate the ratio of length to width to see how close their ratio in decimal form comes to 1.618. You could encourage students to measure some objects in inches and others in centimeters.
3. To add variety to the lab, you could give each group a different size frame or envelope and then compare with other groups to see which particular picture frame or envelope comes closest to the golden ratio.

Possible answers for the lab sheet are as follows:

ARTICLE	LENGTH	WIDTH	RATIO
3 × 5 index card	5 inches	3 inches	1.6667*
5 × 8 index card	8 inches	5 inches	1.6*
Notebook paper	11 inches	8.5 inches	1.294
Picture frame: 2 × 3 3 × 5 4 × 6 5 × 7 8 × 10 10 × 12	 3 inches 5 inches 6 inches 7 inches 10 inches 12 inches	 2 inches 3 inches 4 inches 5 inches 8 inches 10 inches	 1.5 1.6667* 1.5 1.4 1.25 1.2
Playing card (sizes vary)	3.5 inches	2.5 inches	1.4
Envelope (sizes vary)	Business size 9 7/16 inches	4 3/16 inches	2.254

* Closest to golden ratio

Have students discover and share what rectangular shapes in the classroom are close to the golden ratio. Measure the length and width to verify.

Extensions

Because the golden ratio was designed to be pleasing to the eye, you might want to explore the size of various magazine covers or cereal boxes that are designed to catch the eye. There are a variety of Web sites that provide artwork demonstrating the golden ratio. A search with golden ratio provides a variety of sources.

Name________________________________ Date ______________

ACTIVITY 1

The Golden Ratio

The **golden ratio** was first discovered by the ancient Greeks. Many rectangles called **golden rectangles** demonstrate the golden ratio. The ratio of the length of the longer side to the length of the shorter side is called the golden ratio and is approximately 1.618. Because of their pleasing shape, golden rectangles can be found in the shape of playing cards, windows, book covers, file cards, ancient buildings, modern skyscrapers, and art. It is believed by some researchers that classical Greek sculptures of the human body were proportioned so that the ratio of the total height to the height of the navel was the golden ratio.

Directions For the following activity, measure the various articles for length and width and compute the ratio of length to width to 3 decimal places.

ARTICLE	LENGTH	WIDTH	RATIO *(length:width as a decimal)*
small index card			
large index card			
notebook paper			
picture frame			
playing card			
envelope			
top of your desk			

Look at your results in the fourth column. How many of the articles you measured were near the golden ratio?

____

(continued)

Name________________________________ Date ________________

ACTIVITY 1

The Golden Ratio *(continued)*

Look around the room. Look for rectangles that might be golden rectangles, that is, whose ratio of length to width is about 1.6. List them below. With a lab partner, measure your choices and see how close you come to 1.618.

OBJECT SELECTED	LENGTH	WIDTH	RATIO *(length:width as a decimal)*

Activity 2

How Accurate Were Ancient Measuring Devices?

Teacher Page

Learning Outcomes

Students will be able to

- compare ancient measuring devices with those used in the English measuring system.
- look for repeated patterns in the data collected.

Overview

Students will measure various parts of the body using ancient devices and will compare those numbers with various English measurements (e.g., feet and inches).

Time Requirements

45–60 minutes

Group Size

Three or four students per group

Materials

- yardstick and meterstick or measuring tape
- ruler with English and metric measurements
- string

Procedure

1. Divide the class into groups of three or four students and pass out the lab sheet and measuring devices (yardstick, meterstick, ruler, tape measure).
2. For each body measurement, students will
   (a) measure their body part in inches and centimeters.
   (b) compare their results to their teammates'.
   (c) average the team results.
3. Each team will record their averages on the board. The teacher should have the categories displayed so that students can write their results under the proper categories. (See student handout for setup.)

4. Each team will then look at the listed data and look for patterns. Does the human foot equal 12 inches? Is the measure from the tip of the nose to the end of the hand equal to 36 inches? And so on.
5. *Some possible observations:* The closed-palm measurements should not vary too much. This technique is still used today in measuring horses. The nose to tip of hand measurement should come close to 36 inches, depending upon the age of the student. (An adult would come closer than a child to meeting this measure.) The foot measurement should vary owing to differing foot/shoe sizes.

Extensions

Students will discover that the measure of their outstretched arms is equal to their height. With the string/head activity, the string should go around the head approximately three times regardless of how short or tall the person is.

Name____ Date ________________________

ACTIVITY 2

How Accurate Were Ancient Measuring Devices?

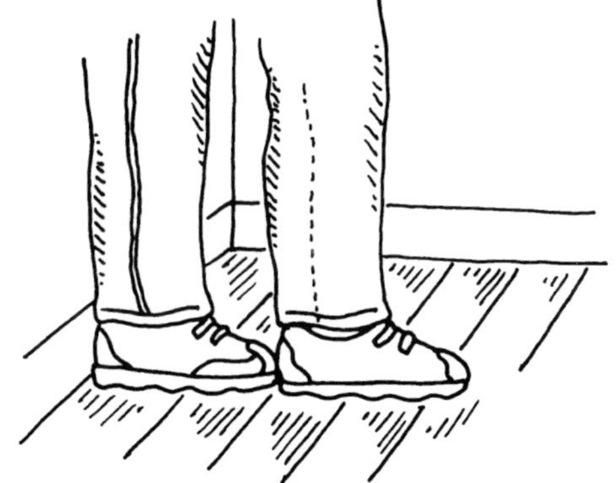

Directions In your team each of you will be asked to measure distances using your hands, feet, and so on. You will then compare your results with those of your teammates. You will record all measurements in inches **and** in centimeters.

Closed palm

1. Measure the width of your closed palm across the widest part of your palm. Record it below.

Measure of closed palm in inches __________ in centimeters __________

2. Compare your results with those of your teammates. Was your measurement equal to, smaller than, or greater than those of your teammates? Discuss.
3. Find the average of the measured palms in your group and record that number below.

in inches __________ in centimeters __________

Outstretched hand

1. Find the measure of your outstretched hand from thumb to little finger and record that number below.

in inches __________ in centimeters __________

2. Compare your results with those of your teammates. Was your measurement equal to, smaller than, or greater than those of your teammates? Discuss.

(continued)

Name____ Date ______________________

Activity 2

How Accurate Were Ancient Measuring Devices? *(continued)*

3. Find the average of the outstretched palms in your group and record that number below.

in inches __________ in centimeters __________

Nose to tip of your hand

1. With your arm extended out from your side, find the measure from the tip of your nose to the tip of your longest finger. You may need to pair up and help each other do this. Record your measurements below.

in inches __________ in centimeters __________

2. Compare your results with those of your teammates. Was your measurement equal to, smaller than, or greater than those of your teammates? Discuss.
3. Find the average of the nose-to-hand measures in your group and record that number below.

in inches __________ in centimeters __________

Foot measure

1. Find the length of your foot with your shoes on and record that measure below.

in inches __________ in centimeters __________

2. Compare your results with those of your teammates. Was your measurement equal to, smaller than, or greater than those of your teammates? Discuss.

(continued)

Name______________________________ Date ______________

ACTIVITY 2

How Accurate Were Ancient Measuring Devices? *(continued)*

3. Find the average of the foot measures in your group and record that number below.

in inches __________ in centimeters __________

Class data

Directions At the front of the room, each team should record their average measures for the four categories above. As the data are presented, record the information below.

TEAM	CLOSED PALM	OUTSTRETCHED PALM	NOSE TO HAND	FOOT

1. What observations can you make based on the data? ______________________________
____
2. Were any measurements almost equal? ______________________________
3. What measurements seem to vary greatly? What might cause this? ______________
____
____
4. Which "human" measuring devices would give you consistent results? __________
____

(continued)

Name____ Date ______________________

Activity 2

How Accurate Were Ancient Measuring Devices? *(continued)*

In ancient times, the length of the human foot evolved into the "12 inches = 1 foot" that we use today.

1. Did the measure of your foot come close to 12 inches? ______________________

   ____

2. How did your group average compare? ______________________________

3. Do you think using your foot would give the same results as using the feet of other classmates? Why or why not? ______________________________________

Extensions

1. It is said that the measure of your outstretched arms from fingertip to fingertip equals your height. Experiment with your teammates to see if this is true.

   Length of your outstretched arms in inches ______________________________

   Your height in inches ____

2. How many times would a string have to go around your head to equal your height? Make a guess, and then see what results you actually get.

   Guess how many times around your head ____________________________

   Actual amount of times around your head ___________________________

   Was the actual amount consistent with your teammates' results?________________

   ____

ACTIVITY 3

Unit Fractions

Teacher Page

Learning Outcomes

Students will be able to

- discover pattern relationships with unit fractions.
- know what a unit fraction is.

Overview

Student will use grid paper to find unit fractions in rectangles and will try to discern patterns and characteristics of unit fractions from a series of rectangles.

Time Requirements

30–45 minutes

Group Size

Either pairs, or groups of three or four students

Materials

- centimeter grid paper
- lab sheet

Procedure

1. Pass out centimeter grid paper.
2. Walk the students through the illustration in the lab. The key to finding the unit fractions in the exercise is always to start with a single square and then try to divide the square by the remaining shaded square parts. It is advisable to demonstrate this activity first before having the students attempt it, to offer help and insights.
3. The chart should be as follows:

RECTANGLE	FRACTION	UNIT FRACTION	+	UNIT FRACTION
2×3	$\frac{1}{2}$	$\frac{1}{6}$	+	$\frac{1}{3}$
3×4	$\frac{1}{3}$	$\frac{1}{12}$	+	$\frac{1}{4}$
4×5	$\frac{1}{4}$	$\frac{1}{20}$	+	$\frac{1}{5}$
5×6	$\frac{1}{5}$	$\frac{1}{30}$	+	$\frac{1}{6}$
6×7	$\frac{1}{6}$	$\frac{1}{42}$	+	$\frac{1}{7}$
7×8	$\frac{1}{7}$	$\frac{1}{56}$	+	$\frac{1}{8}$

4. One pattern they should discover is that the fraction's denominator is the first dimension. Also, the first unit fraction's denominator is the product of the dimensions and the last unit fraction's denominator is the second dimension.

   For example, in a 10 × 11 rectangle, the fraction would be $\frac{1}{10}$ and the sum of the unit fractions would be $\frac{1}{110} + \frac{1}{11}$.

5. Any fraction can be expressed as the sum of unit fractions. You might want to explore why this is so. An example of other fractions would be $\frac{2}{7} = \frac{1}{28} + \frac{1}{4}$. (The dimensions of the rectangle are 7 × 4.)

Name________________________________ Date ________________

ACTIVITY 3

Unit Fractions

The Egyptians developed a way to express fractions as the sum of **unit fractions.** Unit fractions are fractions that have 1 as their numerator. In this lab, you will find the unit fractions by dividing up rectangles.

First you will start with a 2×3 rectangle as shown to the right.

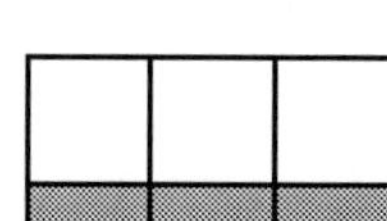

The shaded region represents what fraction? ½

Can you represent ½ as the sum of two unit fractions?

To do this take one square from the rectangle above. What fraction does this square represent? ⅙

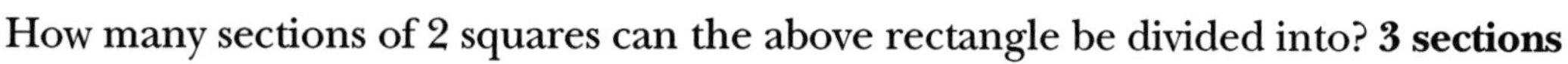

How many sections of 2 squares can the above rectangle be divided into? **3 sections**

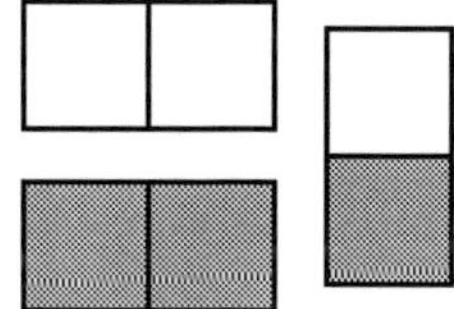

What fraction do any two squares (for instance, the two shaded squares on the left) represent? ⅓

So ½ = ⅙ + ⅓

Now get together with your partner or team. The rectangle's dimensions will be increased by 1. That is, you will look at a rectangle that is 3×4.

To find the unit fractions, do the following:

1. Divide the rectangle into unit squares.
2. Decide what fraction one row of squares represents.
3. Take the first unit square and determine what unit fraction it represents.
4. Determine what unit fraction the rest of the row of squares represents.
5. Also try this for rectangles of 4×5; 5×6; 6×7; 7×8.

(continued)

Name____ Date ____________________

Activity 3

Unit Fractions *(continued)*

You will be given centimeter grid paper to draw the rectangles on. Please fill in the blanks below as you discover the sum of unit fractions for each fraction determined from step #2.

Rectangle	Fraction	Unit Fraction	+	Unit Fraction
2×3	$\frac{1}{2}$	$\frac{1}{6}$	+	$\frac{1}{3}$
3×4	________	________	+	________
4×5	________	________	+	________
5×6	________	________	+	________
6×7	________	________	+	________
7×8	________	________	+	________

What patterns do you notice in representing your fractions as the sum of two unit fractions?

____

____

____

Could you predict how $\frac{1}{8}$ could be expressed as the sum of two unit fractions?

____

____

____

Area

Activity 4

Area/Percent Scavenger Hunt

Teacher Page

Learning Outcomes

Students will be able to

- apply the concept of area with rectangles and other plane figures.
- apply the concepts of ratio, proportion, and percents.
- refine estimation skills.

Overview

Student pairs will estimate areas of rectangles, then compute actual areas and percents using ratio and proportions.

Time Requirements

35–45 minutes

Materials

- meterstick
- calculator
- lab sheet

Group Size

Pairs

Procedure

1. Before beginning the lab, review the formula for area of a rectangle, the meaning of a ratio, and how to calculate percents using proportions.
2. Pass out lab sheets and metersticks.
3. Have students share their findings with the given percents.
4. Next discuss the meaning of percents that are less than a 100% and percents that are greater than a 100%. Have them brainstorm objects outside of the room that might represent these values.
5. Have each team do at least 3 percents greater than 100%, record their results on the chart, and then report to the class.

Extensions

1. Follow the same procedure with other area formulas such as area of a triangle, a square, a circle, and so forth.
2. This activity can also be expanded to include volume formulas such as volume of a cube, a rectangular solid, a sphere, and so forth.

Name____ Date ____________________

Activity 4

Area/Percent Scavenger Hunt

1. With your partner, measure the length and width of a door in meters and record below. (Remember: **area = length × width.**)

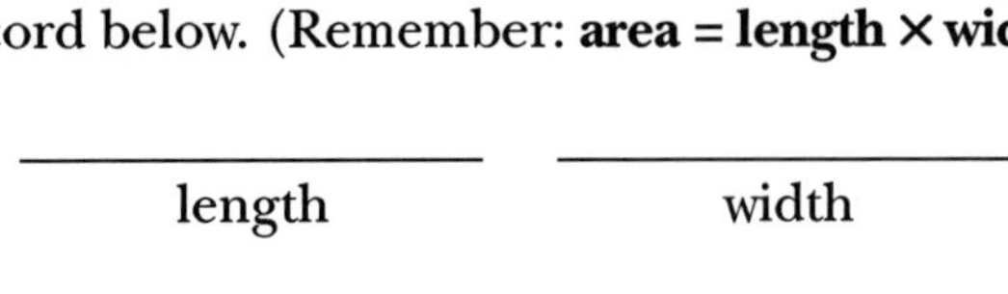

2. Next calculate the area of the door in square meters.

____

3. Look around the room and find an object whose area is about 50% of the area of the door. Write the name of the object and find its length, width, and area.

______________ ______________ ______________

object length width

Calculate the area of this smaller object below.

____

4. Express the ratio of the area of the small object you just calculated to the area of the door.

$\frac{\text{object}}{\text{door}}$ ____

5. Using a proportion, express this ratio as a percent.

$\frac{\text{object}}{\text{door}} = \frac{\blacksquare}{100} = \blacksquare\%$ _______________________________

Compare the percent you found to 50%. How accurate was your prediction?

____

6. Continue with this procedure. Find objects whose areas are 25%, 125%, $33\frac{1}{3}$%, 10%, or 1% of the area of the door.

Remember to
- (a) find the dimensions of the object.
- (b) compute its area.
- (c) make a ratio of the object to the door.
- (d) use a proportion to express this ratio as a percent.

(continued)

Name________________________________ Date ______________

ACTIVITY 4

Area/Percent Scavenger Hunt *(continued)*

7. Fill in the chart below with your results.
8. When you are finished, select four or five more percents greater than 100% and do the same procedures. Work together with your partner.

AREA SCAVENGER HUNT CHART					
ESTIMATED PERCENT	OBJECT	AREA OF OBJECT	AREA OF DOOR	RATIO	ACTUAL PERCENT
50%					
25%					
125%					
10%					
1%					

Activity 5

Discovering the Area of a Circle

Teacher Page

Learning Outcome

Students will be able to derive the formula for the area of a circle through measurement and experimentation.

Overview

Students, using the averaging method, will discover how the formula for the area of a circle was first derived.

Time Requirements

30 minutes

Group Size

Pairs

Materials

- scissors
- scientific calculator
- rulers
- tape
- lab
- Figure *a* and *b* handouts

Procedure

Part One

1. Before beginning this activity, make sure students are familiar with the terms **circumscribe** and **inscribe** and how to find the area of a square.
2. Pass out rulers and lab sheets. Students may measure the sides of the squares in either inches or centimeters. However, centimeters are easier to work with because decimals are involved in the calculations.

Answers

1. The area of the inscribed square EFGH is approximately 8 square centimeters (sq cm).
2. The area of the circumscribed square ABCD is approximately 16 sq cm.
3. The estimated area is $\frac{8 + 16}{2} = 12$ sq cm.
4. The diameter of the circle is 4 cm.
5. The radius of the circle is 2 cm.
6. The area of the circle is 12.56 sq cm.

7. In comparing, the averaging method was off by 0.56 sq cm. The teacher might want to show how small a square centimeter is and that the margin of error using averaging was about half that.

Part Two

1. Pass out scissors, tape, and two copies each of Figure *a* and Figure *b*. (One copy will be used to compare the cutouts to the original.)
2. Students should cut out the larger square in Figure *a* first. Students should leave the second copy of Figure *a* intact to compare when answering the questions. Then they should cut out the four smaller squares that form the large square.
3. The dimensions of the smaller squares are $r \times r$.
4. The area of one of the smaller squares is r^2.
5. The area of the larger square is $4r^2$.
6. Next have students cut out the smaller square in Figure *b* leaving another Figure *b* intact to compare with when answering the questions. After students cut out the smaller square, they should cut that square into four smaller triangles following the lines provided. They should have four right triangles whose height is *r* and whose base is *r*.
7. Two of the right triangles will form a square whose dimensions will be $r \times r$. Therefore, they will be able to form two smaller squares. Have them tape these squares on their lab sheet as a visual.
8. The dimensions of one of the smaller squares is $r \times r$.
9. The area of one of the smaller squares is r^2.
10. The area of the inscribed square is $r^2 + r^2 = 2\,r^2$.
11. $4\,r^2 + 2\,r^2$ divided by $2 = 3\,r^2$
12. The area of a circle is approximately $3.14\,r^2$; the averaging method comes very close to this with $3\,r^2$.

Name____ Date ______________________

Activity 5

Discovering the Area of a Circle

Part One

The Averaging Method

One method of finding the approximate area of a circle is by using the **averaging method.** Because circles are curved, ancient mathematicians used **inscribed** and **circumscribed** regular polygonal shapes in and around circles. By averaging areas of the two polygonal shapes, the mathematicians came very close to the area of a circle. In this activity, we will find the area of the circumscribed square *ABCD (the square that surrounds the circle)* and find the area of the inscribed square *EFGH (the square that is inside the circle).*

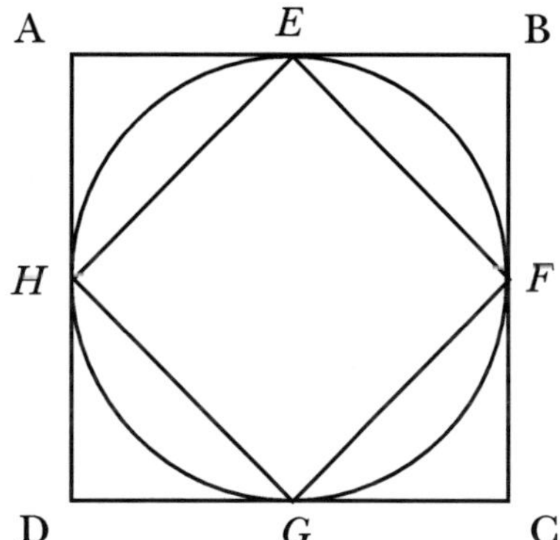

1. The area of the inscribed square EFGH is ________________.
2. The area of the circumscribed square ABCD is ________________.
3. Therefore, the estimated area of the circle found by averaging is ____________.

   Now we will compare our estimated area to actual measurements of the circle, using the formula for area of a circle. *(Use the π key on your calculator.)*

4. The **diameter** of the circle is ______________.
5. The **radius** of the circle is ________________.
6. The area of the circle using the formula π times the radius squared is _____________.
7. Compare the averaging method to the actual formula. How do they compare?____

(continued)

Name____ Date ____________________

ACTIVITY 5

Discovering the Area of a Circle *(continued)*

Part Two

Finding the Area of a Circle Abstractly

To prove the area formula abstractly, we will now use the following diagrams.

1. Cut out both figures *a* and *b*.
2. In Figure *a,* cut out the four smaller squares that form the larger square.
3. What are the dimensions of the smaller squares?__________ × __________
4. What is the area of one of the smaller squares of Figure *a?*__________
5. Because the larger square is made up of four smaller squares, what is the area of the larger square?

Figure a

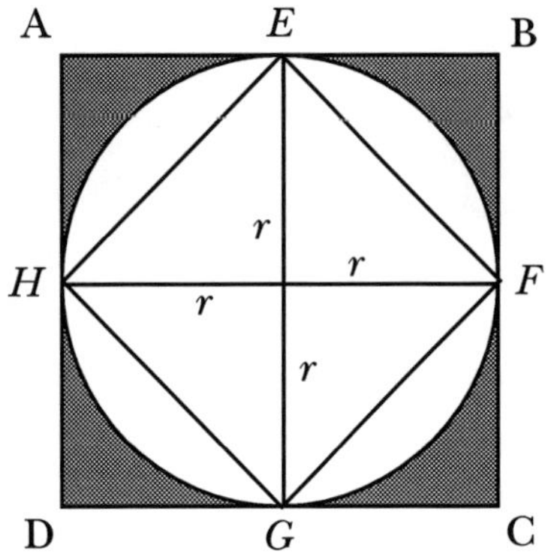

A E B H F D G C r r r r

Figure b

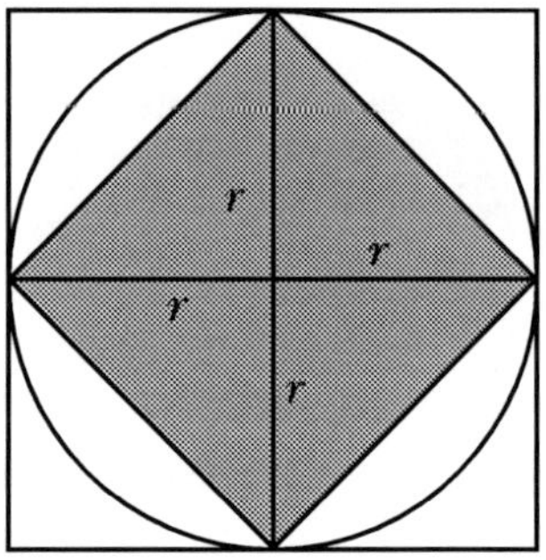

r r r r

(continued)

Name____ Date ____________________

ACTIVITY 5

Discovering the Area of a Circle *(continued)*

6. In Figure *b*, cut out the four smaller triangles.
7. How many little squares can you form with the four pieces? ________
8. What are the dimensions of one of the squares formed? ________ × ________
9. What is the area of one of the squares formed? ________
10. What is the area of the entire square that is *inscribed* (inside) in the circle?

    ____

11. Using the averaging method, average the areas you found above.

    (________________ + ________________) ÷ 2 = ____________________
    area of larger square   area of smaller square   approximate area of circle

12. The area of a circle has been defined as πr^2 or $3.14r^2$. Does the approximate method give a result close to this?

    ____

Activity 6

Circumference vs. Area of a Circle

Teacher Page

Learning Outcomes

Students will be able to

- apply the concept of circumference and area with circles.
- discover the effect on the area of doubling the circumference.
- discover the effect on circumference and area of changing a diameter.

Overview

By experimenting with several circles, students will discover that doubling the circumference quadruples the area.

Time Requirements

45–60 minutes

Group Size

Pairs

Materials

- circular objects of various sizes
- calculator
- lab sheet
- centimeter cube, or centimeter square made from posterboard
- plain paper
- centimeter-grid paper
- scissors
- string

Procedure

1. Have a selection of circular objects in various sizes available, such as jars, lids, paper dishes, and cups. Or you can cut circles of various sizes out of posterboard ahead of time. Centimeter cubes are commercially available. If you do not have them, posterboard can be used.
2. Each pair will select two different circular objects, and each partner will independently estimate the number of square centimeters in the surface of each of the two circular objects, using a centimeter cube or centimeter-square cardboard model. They will compare results and come to consensus on an estimate.
3. Next, each partner will trace the two circles on centimeter-grid paper and estimate the square area by counting the number of square centimeters within the surface of the circles. Next they will compare their estimates. These should be within a few counts of each other.

4. The students will discuss the two methods of estimating and decide which one they feel is most accurate. They will then record agreed-upon results.
5. The students will take string and wrap it around the circumference of each circular shape and then cut it to the length of each circumference.
6. Have students take the length of doubled string of each circumference and form a new circle on the table, predicting what they think the estimated surface area of each will be. Then they should form the circles on centimeter-grid paper, tape it down securely and estimate again, counting the number of squares within each new circle formed. (This is where you may have to instruct them to tape two 8½" × 11" sheets of grid paper together.)
7. Determining the circumference using string and then doubling that string to form a new circle results in students' thinking the area will also be doubled. When they form a new circle on centimeter-grid paper and estimate the number of squares, they should begin to discover that if the radius of a circle is doubled, the new circumference is doubled ($2[\pi 2r]$). And the new area is quadrupled ($\pi [2r]^2 = 4\pi r^2$).

Name______________________________ Date ____________________

ACTIVITY 6

Circumference vs. Area of a Circle

1. With your partner, select two different small circular objects, such as a can, jar, lid, or dish, and trace the outline on plain paper. Next, take a centimeter cube and estimate the number of square centimeters in the surface of your circular objects. Each of you should estimate both objects and compare your results. Circle *a* is a ______________ with an estimated ________________ sq cm, and Circle *b* is a ______________ with an estimated ______________ sq cm.

2. Next, using centimeter-grid paper, trace around your objects a second time and determine the area by counting the number of square centimeters. Circle *a* is __________________ sq cm and Circle *b* is _________________________ sq cm.

3. Compare the two techniques of estimating for each circle and determine which technique is more accurate, and record it.

   Circle *a* has approximately ___________________ sq cm, and Circle *b* has approximately ___________________ sq cm.

4. Now take a piece of string, wrap it around the rim or **circumference** of your object, and cut it the length of the circumference. Measure it in centimeters, then cut a second piece of string that is double the length of the first. The length of the circumference is ______________ cm. Doubling the circumference results in a length of ______________ cm.

5. If you form the doubled string into a circle, what do you predict the new area will be? Write your estimate here. ____________________ sq cm. To confirm your estimate, form the circle on a piece of centimeter-squared paper. (**Note:** You will have to tape two $8\frac{1}{2}$" × 11" pieces together to get a length of 22".) Tape it down. Estimate the area by counting the number of square centimeters. Write your prediction for the area of the larger circle here. ___________ sq cm.

6. Was your first visual estimate close to this second estimate? If not, how do you explain the differences?

____

____

(continued)

Name____ Date ____________________

ACTIVITY 6

Circumference vs. Area of a Circle *(continued)*

7. Using the data from your experiments, how does doubling the circumference affect the area? ____

____

____

____

____

____

8. Using the two circle formulas of circumference and area, conclude by writing a mathematical statement that proves your discovery.

____

____

____

____

____

9. Explain how pizza companies could use this information to their advantage.

____

____

____

____

____

Activity 7

Area of a Parallelogram and a Trapezoid

Teacher Page

Learning Outcomes

Students will be able to

- develop the formulas for the area of a parallelogram and the area of a trapezoid.
- identify the base and height for a parallelogram and trapezoid.
- understand the concept of area as square units.

Overview

Students will work with cutout trapezoids and parallelograms to discover area formulas.

Time Requirements

30 minutes

Group Size

Pairs

Materials

- parallelogram and trapezoid shapes (see page 37)
- scissors
- tape

Procedure

Part One

Area of a Parallelogram

1. Distribute photocopied parallelograms and trapezoids from page 37 for each pair of students. Each student should have one parallelogram and two identical trapezoids to cut out.
2. Make sure the students label the base and the height of the cutout parallelogram correctly before they begin to separate the parallelogram along its height.

Teacher Page

3. Some students may need help in turning their pieces to form a rectangle. The cut pieces will form a rectangle as shown below.

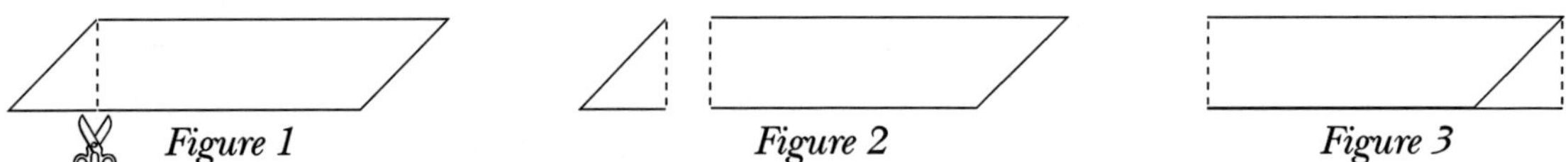

Figure 1 *Figure 2* *Figure 3*

Answers

4. Base times height or $b \times h$
5. Yes, because the base and height are the same for the original parallelogram as they are for the newly formed rectangle.
6. A = bh

Part Two

Area of a Trapezoid

1. Pass out the photocopied trapezoids. Each student should receive two identical trapezoids to cut out.
2. Make sure the students have correctly labeled the parallel bases as well as the height.
3. The two trapezoids will form a large parallelogram as shown below.

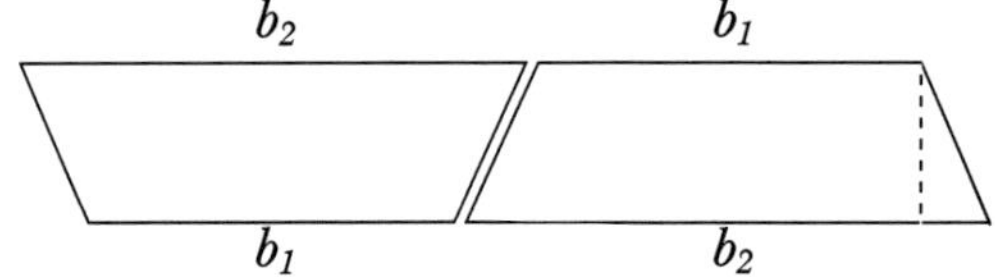

Answers

4. $b_1 + b_2$
5. h
6. **A** = bh
7. $A = (b_1 + b_2)h$
8. $A = \frac{1}{2}(b_1 + b_2)h$

Name______________________________ Date ______________

ACTIVITY 7

Area of a Parallelogram and a Trapezoid

Part One

Area of a Parallelogram

You and your partner will each be given a **parallelogram** that has the height shown by a dotted line and the base as the side the height is drawn perpendicular to. Cut out each shape and label as follows:

1. Label the base ***b*** and the height ***h*** of your parallelogram.
2. Carefully separate the parallelogram by cutting along the dotted line, the *height.*
3. Rearrange the two pieces to form a rectangle. Tape the newly formed rectangle below. Then outside the rectangle, label the length as ***b*** and the width as ***h.***

4. How can you find the area of the rectangle you have just formed? ______________

____

5. Will the area of the original parallelogram be the same as the area of the rectangle you just formed? Explain.______________________________

____

6. What do you think might be a formula for finding the area of a parallelogram? Write your formula. ______________________________

(continued)

Name____ Date ____________________

ACTIVITY 7

Area of a Parallelogram and a Trapezoid *(continued)*

Part Two

Area of a Trapezoid

You and your partner will each be given identical **trapezoids.** Cut out each shape and label as follows:

1. Label the shorter parallel side b_1 and the longer parallel side b_2.
2. Label the height in each of the trapezoids as ***h***. Remember the height is always drawn from one base **perpendicular** to the other base.
3. Arrange the two trapezoids to form a parallelogram. Tape the pieces below.
4. For this parallelogram you have just formed, how would you express the base of this parallelogram in terms of b_1 and b_2? ____________________
5. What is the height of this parallelogram? ____________________
6. What is the area for any parallelogram? ____________________
7. Using the base from #4 and the height from #5, what is the area of your parallelogram shape? ____________________
8. Because the formula in #7 finds the area of two trapezoids together, discuss with your partner what the area of one trapezoid would be and write your answer here.

____

(continued)

Name______________________________ Date ______________

ACTIVITY 7

Area of a Parallelogram and a Trapezoid *(continued)*

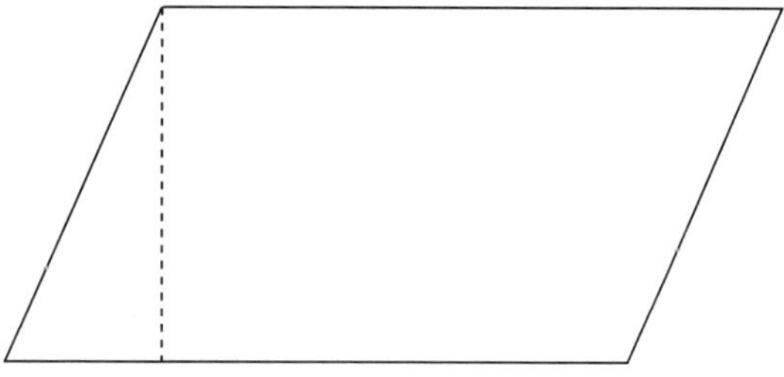

Activity 8

Surface Area of a Sphere

Teacher Page

Learning Outcomes

Students will be able to

- discover the formula for the surface area of a sphere.
- understand the difference between surface area and volume.

Overview

Students, using string and Styrofoam™ balls, will discover the surface area formula for spheres.

Time Requirements

30 minutes

Group Size

Pairs

Materials

- Styrofoam™ balls cut in half
- push pins (two per team)
- string that doesn't stretch

Procedure

1. Distribute one half of a Styrofoam™ ball (baseball size), two push pins, and about a foot of string to each team.
2. The flat part of the sphere is in the shape of the circle. The students should place the push pin in the center of this circle. The teacher should check each team to make sure this is done.
3. Students begin to wind the string around the pin until the entire flat surface is covered, then cut off any excess string.
4. The formula they should recognize for the area the string covers is the area of a circle (πr^2).
5. Next, have students unwind the string, then measure and cut another piece of string the same length.

Teacher Page

6. The next step is for the students to place the push pin in the center top of the curved region.

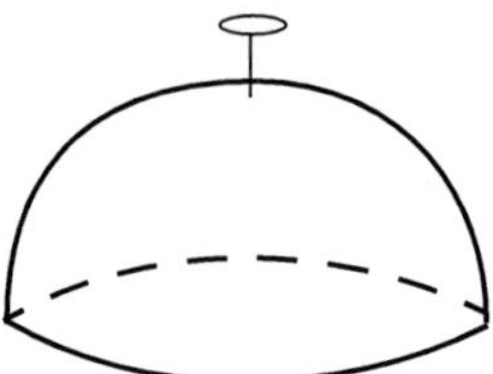

7. Once the pin is placed, students wind the string around the pin until it is used up. The second piece of string is then attached and students continue to wind the string until the entire surface of the half-sphere is covered. If done correctly, the two strings should cover the entire half-sphere.
8. The area that one string covered is πr^2; therefore, two strings cover twice that area, or $2\pi r^2$.
9. Therefore, the surface area of a sphere (two half-spheres) is double this, or $4\pi r^2$.

Name______________________________ Date ______________

Activity 8

Surface Area of a Sphere

You and your partner will be given half a Styrofoam™ ball, two push pins, and string.

What shape is the flat part of the **sphere?** ______________________

Place a push pin in the center of the flat region and wrap string around the pin until the entire flat circular region is covered.

What formula represents the area you covered with the string? ____________

Unwrap the string. Measure and cut a second piece of string the same length.

Next put the push pin in the center of the curved surface and start to wrap the string around the curved surface. When the string is used up, put in a second push pin to hold the string on the curved surface. At the same time, attach the second string where the first one ended and continue to wrap the string around until the entire curved surface is covered.

Did you use the entire amount of cut string? ____________________

The area that one string covered was ______________________ .

Therefore, the area that two strings covered is ____________________. This is also the **surface area** for half a sphere.

With that information complete the following statement:

The surface area of a sphere is ________________.

(continued)

Angles

Activity 9

What's My Angle?

Teacher Page

Learning Outcomes

Students will be able to

- label and describe angles four different ways.
- use patterns to find a formula.

Overview

Students will pair up with outstretched arms to form angles and name them. They will also work on paper to discover a pattern in the number of angles created each time a ray is added.

Time Requirements

30–45 minutes

Group Size

Pairs (formed *after* students receive slips of paper)

Materials

- slips of paper with letters (*A, B, B, C* for one set; *D, E, E, F* for another set; and so on)

Procedure

Part One

1. Give each student two letters, one of the duplicate letters and another single letter. For example, one student would receive the letters *A* and *B;* another student would receive *B* and *C.*
2. Have students look for their matching letter and join that student to form an angle. By physically forming an angle, students will understand both the importance of order in describing an angle and the formation of an angle. That is, that the vertex (here, the duplicate letter) is always in the middle and the other two letters describe how the angle is drawn. For example, angle *ABC* could be read angle *ABC,* or angle *CBA,* or angle *B,* but not angle *BCA.*
3. Have the students read their angle out loud to you (for example: "Angle *GHI*").

Part Two

1. You can choose to talk about acute, obtuse, straight, and right angles before this activity, or you can use this activity as a discovery approach so students can guess what an acute angle is. If you choose the latter approach, you would say which groups have formed an acute angle and which groups have not. Through the process of elimination, the students should be able to see the characteristics of an acute angle.
2. Have students form their angles at different locations in the classroom. You will say "Form an acute angle" and the teammates must position themselves to look like an acute angle.
3. Proceed in this way with obtuse angles, straight angles, and right angles.

Part Three

In this activity, students are to discover a pattern with the angles created each time a ray is added. The chart below shows the values. Students will see that the number of angles increases sequentially.

Number of Rays	Number of Angles
0	1
1	3
2	6
3	10
4	15
5	21

Name____ Date ____________________

ACTIVITY 9

What's My Angle?

Part One

You will be given two letters. Once everyone has his or her letters, try to find someone in your class who has one of the same letters as you. Together you will form an angle by touching hands holding the letter you share. Your other arm, along with your partner's, will form the sides of the angle. Say the name of your angle out loud and then write the name of your angle below. Remember there are three ways to name an angle.

∠ ______________ ∠ ______________ ∠ ______________

Part Two

Angles are classified into four groups: **acute, right, obtuse,** and **straight** angles. Your teacher will ask you to form one of these angles. See if you can guess what the angle will look like. Watch the other groups as well. Your teacher will tell you if you have made the correct angle. Once you have made the correct angle, write a description of each angle below.

Acute angle ____

____

Right angle ____

____

Obtuse angle ____

____

Straight angle ____

____

(continued)

Name____ Date ______________________

Activity 9

What's My Angle? *(continued)*

Part Three

You will try to discover a pattern with angles. We will start with one angle. Then we will add one **ray** into the angle. How many angles does this form? Continue with this pattern, adding another ray and counting how many angles are formed. Can you come up with a formula for figuring out how many angles there are, given the number of rays added?

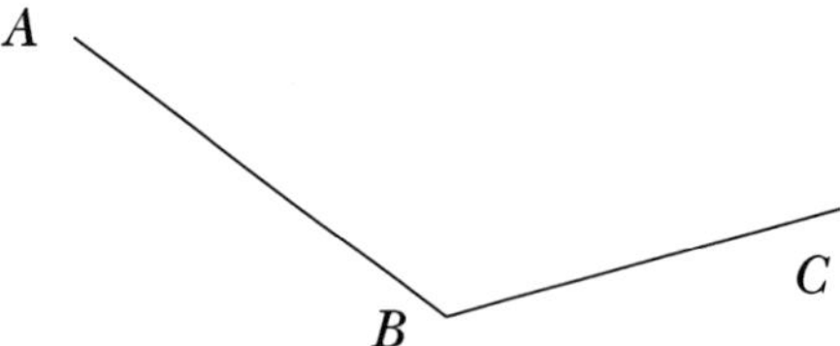

no rays, 1 angle

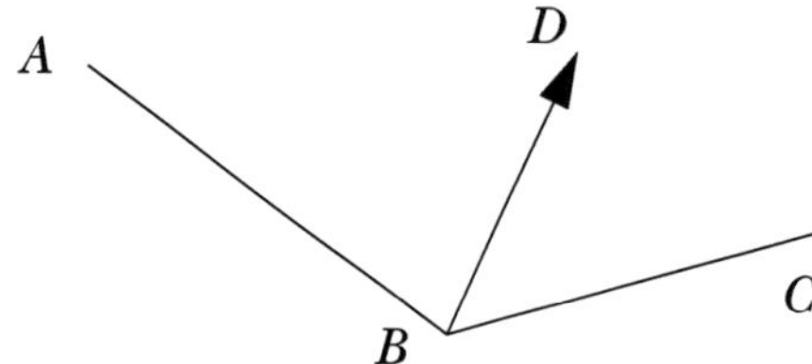

1 ray, 3 angles
($\angle ABD$, $\angle DBC$, $\angle ABC$)

Continue with this pattern. Then fill in the chart on page page 49.

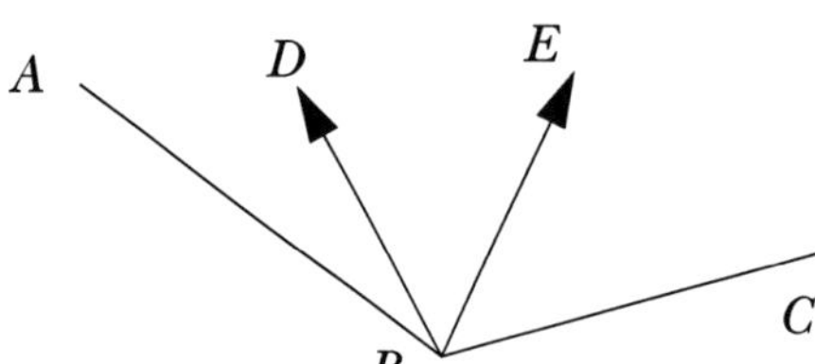

2 rays, ________ angles

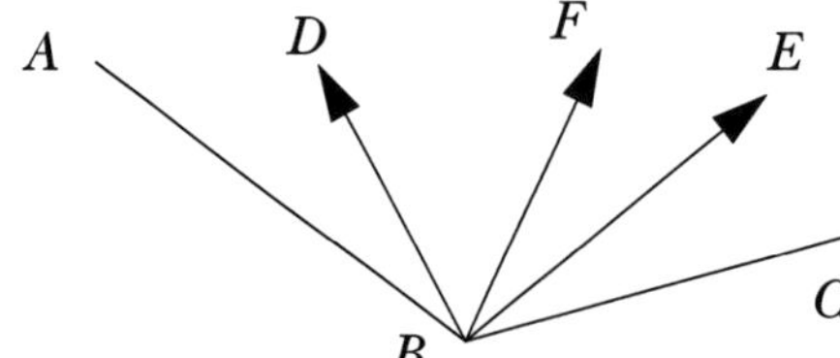

3 rays, ________ angles

(continued)

Name________________________________ Date ________________

ACTIVITY 9

What's My Angle? *(continued)*

NUMBER OF RAYS	NUMBER OF ANGLES
0	1
1	3
2	
3	
4	
5	

Can you find a method for determining how many angles are formed when you are given the number of rays? Describe it below and write it as a formula.

____

____

____

____

____

____

____

____

____

____

ACTIVITY 10

How Many Degrees?

Teacher Page

Learning Outcomes

Students will be able to

- recognize a pattern involving different-shaped polygons and their total angle measures.
- develop a formula for computing the total number of degrees for any convex polygon, where no diagonals lie outside the polygon.

Overview

Students will draw convex polygons of 5 to 10 sides, divide them into triangles, and compute the number of degrees in each polygon. They will then try to discern a pattern in the exercise and generate a formula for finding the number of degrees in any polygon.

Time Requirements

30 minutes or more

Group Size

No groups

Materials

- lab sheets
- rulers
- protractors

Procedure

1. Before students start drawing various-sided polygons, you might want to explain what a convex polygon is and how to divide the region into triangles by drawing diagonals from one vertex to opposite vertices.
2. As students begin to draw their various polygons, make sure that the polygons really are convex or the method will not work.

3. When complete, the chart should be filled in as shown below.

NUMBER OF SIDES	NUMBER OF TRIANGLES	TOTAL DEGREES
3	1	180
4	2	360
5	3	540
6	4	720
7	5	900
8	6	1,080
9	7	1,260
10	8	1,440

4. Students should notice the following:
   (a) The number of triangles drawn is always 2 less than the number of sides in the polygon.
   (b) The total number of degrees increases by 180 each time.
   (c) The strategy for finding the total number of degrees for any convex polygon would be to take the number of sides, subtract two, and then multiply this number by 180°. As a formula, it would be: **$(n–2) \bullet 180°$** where n = the number of sides of thc polygon.

Name________________________________ Date ________________

ACTIVITY 10

How Many Degrees?

The smallest enclosed figure is the triangle, with 180°. As the number of sides of a **polygon** increases, how many triangles can you draw within the figure? What is the sum of the angles? To experiment with this question, the figures drawn must be **convex** polygons. That is, all the sides push out, not in.

convex polygon

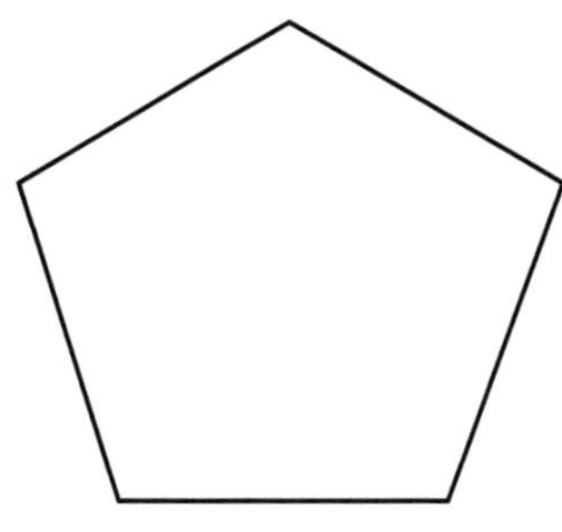

polygon that is not *convex*

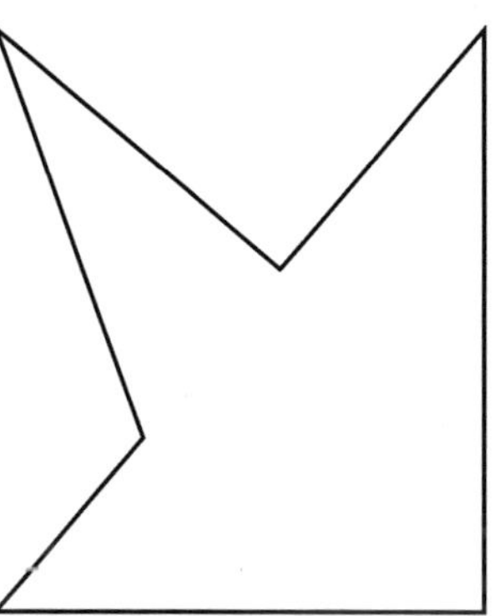

Example: In a four-sided convex polygon, two triangles can be drawn.

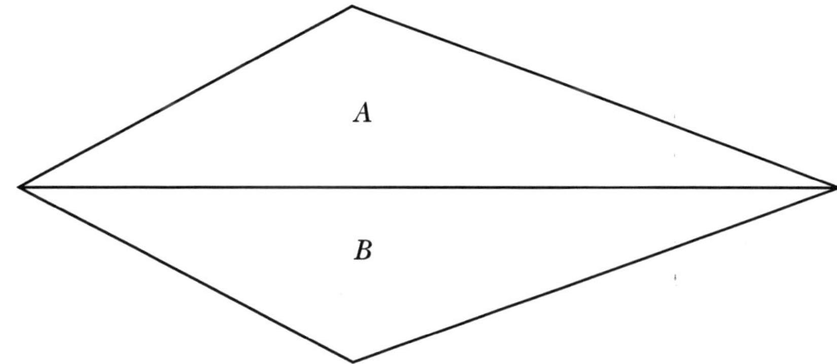

Therefore, you can find the sum of the angles by adding 180° from triangle *A* to 180° from triangle *B,* for a total of 360° for the four-sided figure.

(continued)

Name____ Date ____________________

Activity 10

How Many Degrees? *(continued)*

Continue this process for a five-sided, six-sided, seven-sided figure, and so on, up to a ten-sided figure. First, draw the figure with the correct number of sides. (Make sure all sides push out.) Then, from one **vertex,** draw diagonals to the opposite vertices. Fill in the chart that follows the number of triangles drawn and the total number of degrees in each.

1. Five-sided polygon	4. Eight-sided polygon
2. Six-sided polygon	5. Nine-sided polygon
3. Seven-sided polygon	6. Ten-sided polygon

(continued)

Name______________________________ Date ______________

ACTIVITY 10

How Many Degrees? *(continued)*

NUMBER OF SIDES	NUMBER OF TRIANGLES	TOTAL DEGREES
3		
4		
5		
6		
7		
8		
9		
10		

Do you notice any patterns? ______________________________

____

____

____

____

Can you determine a generalization or formula for determining the number of degrees for any polygon? ______________________________

____

____

____

____

____

ACTIVITY 11

Properties of Parallel Lines

Teacher Page

Learning Outcomes

Students will be able to

- discover that alternate interior angles are equal.
- discover that the perpendicular distance is always the shortest between two parallel lines.
- apply the properties of parallel lines to real-life situations.

Overview

Student pairs will discover various congruencies of angles formed by a transversal intersecting parallel lines.

Time Requirements

35–45 minutes

Group Size

Pairs

Materials

- ruler
- protractor
- lab sheet
- ruled paper

Procedure

Part One

1. If students are not skilled using a protractor, this must be demonstrated first.
2. Model how to darken two lines on a ruled piece of paper and then draw a random line, a transversal, intersecting the two parallel lines at an angle other than 90°.

   Demonstrate how to number the angles consecutively following the model provided in the lab.
3. As students compare their measurements, the teacher should define the special angles of **vertical, corresponding, alternate interior,** and **alternate exterior** as described below.

   Student measurements will differ, but the following congruencies will occur:

   (a) Angles *1* and *2; 3* and *4; 5* and *6; 7* and *8; 2* and *4; 6* and *8; 1* and *3;* and *5* and *7* are supplementary (that is, the sum of the angles adds up to 180°).

   (b) Angles *1* and *4; 2* and *3; 5* and *8; 6* and *7* are congruent and are vertical angles.

(c) Angles *1* and *5; 2* and *6; 3* and *7; 4* and *8* are congruent and are corresponding angles.

(d) Angles *4* and *5;* and *3* and *6* are congruent and are alternate interior angles.

(e) Angles *2* and *7;* and *1* and *8* are congruent and are alternate exterior angles.

(f) Angles *3* and *5;* and *4* and *6* are supplementary.

Part Two

The purpose of this activity is for students to discover that the perpendicular distance is always the shortest distance between two parallel lines. As students compare their transversal segments to the perpendicular segment they drew, they will see that the measures of these transversal segments are always larger.

Part Three

Some possible answers would be: railroad tracks and street intersections.

Encourage students to find as many angle relationships as they can from their example.

This will reinforce the special angles discovered in Part One.

Name____ Date ____________________

ACTIVITY 11

Properties of Parallel Lines

Part One

1. With your partner, take two pieces of ruled paper. Then, working on your own and using a straightedge, each of you darken a different set of parallel lines on your own paper. Make sure the distance between the two lines you have darkened is different from the distance between your partner's darkened lines.
2. Next draw a ***transversal,*** a line that intersects the two parallel lines you have darkened on your ruled paper.
3. Label each of your angles with a number 1 through 8 like the diagram below.

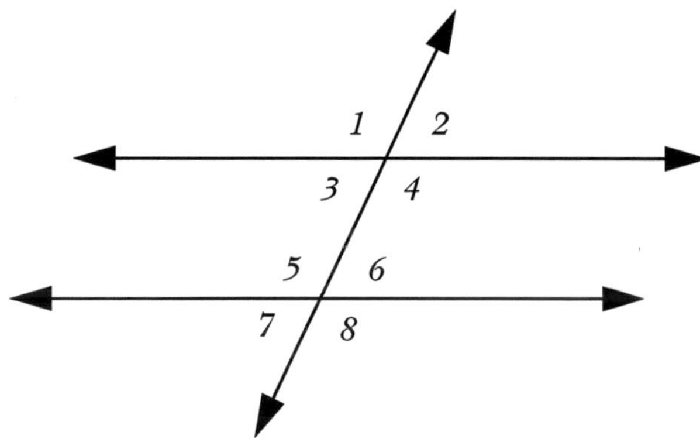

4. Now use a protractor to measure each angle and record the results. Compare your diagram and angles with your partner's.

Diagram 1

$\angle 1$ ___________	$\angle 3$ ___________	$\angle 5$ ___________	$\angle 7$ ___________
$\angle 2$ ___________	$\angle 4$ ___________	$\angle 6$ ___________	$\angle 8$ ___________

(continued)

Name____ Date ____________________

ACTIVITY 11

Properties of Parallel Lines *(continued)*

5. Repeat this experiment two more times by darkening different pairs of lines on your ruled paper. Draw transversals that intersect your parallel lines at different angles. Label and measure each of your eight angles for each new diagram and record below.

Diagram 2

$\angle 1$ __________ $\angle 3$ __________ $\angle 5$ __________ $\angle 7$ __________

$\angle 2$ __________ $\angle 4$ __________ $\angle 6$ __________ $\angle 8$ __________

Diagram 3

$\angle 1$ __________ $\angle 3$ __________ $\angle 5$ __________ $\angle 7$ __________

$\angle 2$ __________ $\angle 4$ __________ $\angle 6$ __________ $\angle 8$ __________

(a) List all pairs of **supplementary angles.** ____________________________

____

____

____

(b) List all pairs of angles that are equal.

____

____

____

Part Two

1. Using the two parallel lines in Diagram 4, draw a transversal that is perpendicular to the lines. Use your protractor to make sure the line you draw is perpendicular or makes 90° angles with your parallel lines. With your ruler measure the **line segment** on that transversal between the two lines and record its measure in centimeters here. ____

(continued)

Name______________________________ Date ______________

ACTIVITY 11

Properties of Parallel Lines *(continued)*

Diagram 4

2. Now draw three more transversals that are *not* perpendicular on the same diagram above and measure the segments. Record below.

______________	______________	______________
Segment 1	Segment 2	Segment 3

Compare your lengths with your partner's. What have you discovered?

____

____

____

____

Part Three

Visualize a place in your community that would be an example of two parallel lines being cut by a transversal. Draw a sketch below labeling all the angles and describing all the math that is related to your sketch.

Activity 12

Angles Inscribed in Circles

Teacher Page

Learning Outcomes

Students will be able to

- measure various angles formed by triangles whose vertex is a point on the circle and whose two vertices are endpoints of the diameter.
- discover that any angle inscribed in a semicircle is a right angle.

Overview

Students will construct various triangles on a circle with vertices at the endpoints of the diameter and discover that the triangles formed are always right angles.

Time Requirements

30–45 minutes

Group Size

Pairs

Materials

- construction paper 14" × 18"
- compass
- protractor
- ruler
- scissors

Procedure

1. Distribute compasses, construction paper, protractors, scissors, and rulers to each pair.
2. Teacher may need to review terms: diameter, radius, circumference.
3. Teacher may need to review how to draw a circle with a compass. The circles for this lab should have a radius of 3".
4. Once the circles are drawn, have the students cut out and fold along the center to form the diameter.
5. Have students locate point C randomly on the circle similar to Figure 1 provided in the lab. They will connect this point to the endpoints of the diameter forming triangle ABC.
6. Each time they measure $\angle ACB$ they should discover that $\angle ACB$ is a right angle. This will occur at any point on the circle provided the angle is drawn to the endpoints of the diameter.

Teacher Page

Answers

6. Angle *ACB* is always a right angle.
7. $\angle BAC$ and $\angle CBA$ are always acute angles and complementary.
8. Any angle formed by a point on the circle and two endpoints of the diameter is always a right angle.

Name____ Date ______________________

ACTIVITY 12

Angles Inscribed in Circles

1. Using a compass, you and your partner each construct a circle with a radius of 3" drawn on construction paper whose dimensions are 14" × 18".
2. Next, each of you cut out your circles, labeling the center point with a black dot. Fold the circle in half, making a crease through the center point. This crease represents the ______________________ of your circle.

   Label the points on either end of the crease A and B. Your circle should look like Figure 1.

Figure 1

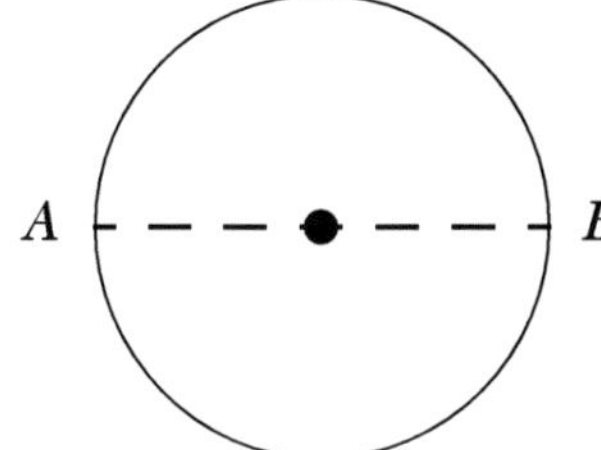

Darken a third point C on the circumference of the circle so that the **endpoints** of the diameter $\overline{AB}$ join with point C creating a triangle within your circle.

Figure 2

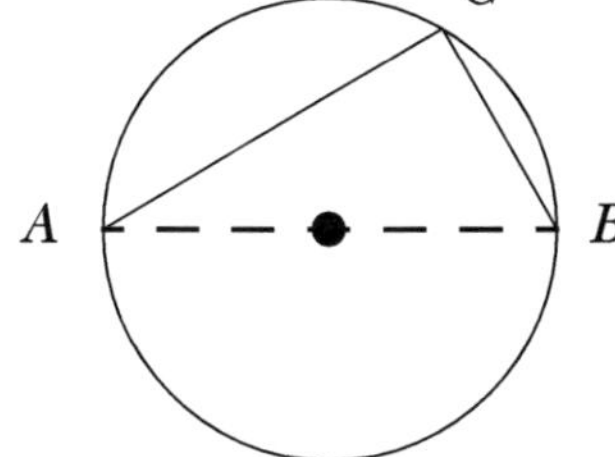

3. Using a protractor, measure angle *ACB*.

   ∠*ACB* ____________

(continued)

Name____ Date ____________________

Activity 12

Angles Inscribed in Circles *(continued)*

4. Now move point C to another position on the same circle and form angle ACB again. Measure $\angle ACB$ is ____________.

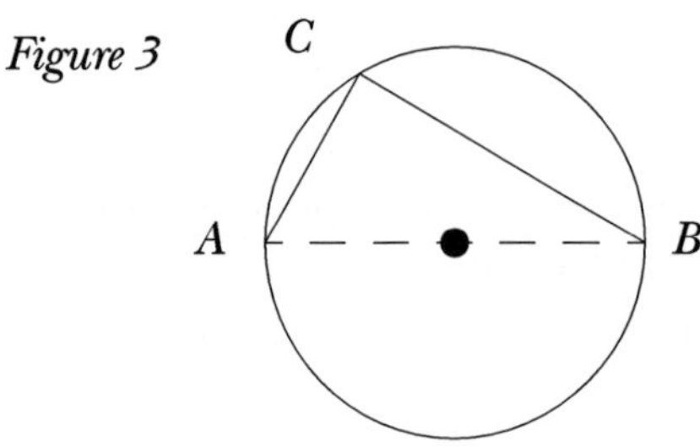

Figure 3

5. Repeat this experiment two more times, rotating point C to various locations on the circle. Draw a diagram of your circles below.

6. What type of angle is $\angle ACB$ in every situation?

____

7. What type of angles were $\angle BAC$ and $\angle CBA$ in every situation?

____

8. What statement can you make about any angle whose vertex is a point on the circle and whose endpoints are the endpoints of the diameter?

____

____

Activity 13

Sums of Angles in a Triangle

Teacher Page

Learning Outcomes

Students will be able to

- construct acute and obtuse angles within a triangle.
- measure angles concisely using a protractor.
- discover that it is impossible to construct a triangle with two obtuse angles.
- discover that the sum of the measures of the angles in any triangle always equals 180 degrees.

Overview

Through angle construction and manipulation, students will discover that the sum of the measures of the angles of a triangle always equals 180 degrees.

Time Requirements

45 minutes

Group Size

Pairs

Materials

- lab sheet
- protractor
- ruler
- construction paper
- highlighters
- scissors

Procedure

Before beginning the lab, you should review the definitions of acute, obtuse, straight, and right angles. (An acute angle is an angle less than 90 degrees, and an obtuse angle is one greater than 90 degrees but less than 180 degrees. A straight angle has 180 degrees, and a right angle has 90 degrees.) The skill of measuring angles with a protractor should also be reviewed.

1. Allow students plenty of time to draw their triangles with one obtuse and one acute angle. Students often cannot figure out how to draw the obtuse angle because their natural inclination is to draw an acute angle first. If they draw an acute angle first, then they will have a triangle with three acute angles. They must draw an obtuse first. If you need to give them hints, model manipulating two rulers showing the direction the side has to be drawn (out from the vertex, not in).

2. Next have students measure the angles to determine if they have followed the criteria, record the angle measures within, and then use a highlighter to color the corners of their triangle.

3. Have students discuss with their partners if it is possible to construct a triangle with more than two acute angles. Since it is possible, encourage them to construct an example on the lab sheet, having them also measure the angles to confirm their findings.

4. At this point they have constructed many triangles, so they have probably figured out that it is impossible to draw a triangle with two obtuse angles. Don't tell them yet that a triangle can have a total of only 180 degrees and that two obtuse angles within one triangle would exceed this value. By continuing with Steps 5–8 students can discover this theorem.

5.–7. As students continue the lab, you may want to have the class do these steps together with you, using a large triangle in the front of the class as a model.

8. Give students ample time to talk with their partners about their discoveries. Have them write their responses on the lab sheet and then bring the groups together for a class discussion.

Name______________________________ Date ______________

ACTIVITY 13

Sums of Angles in a Triangle

1. Using your protractor and ruler, you and your partner each draw a triangle with one **acute angle** and one **obtuse angle.** After you draw your own triangle, measure the angles to confirm that you have one acute and one obtuse angle. Write the angle measures on the triangle. Swap triangles and check your partner's angle measures. Write your angle measures below.

   Acute angle measures __________ degrees.

   Obtuse angle measures __________ degrees.

   The third angle in the triangle is an __________ angle with a measure of __________ degrees.

2. Cut out your triangle. Color each corner of your triangle and record the angle measure in degrees within each colored region. Set aside your triangle to be used again in Steps 5–8.

3. With your partner, discuss the possibility of drawing a triangle with more than two acute angles. Is this possible? __________ Explain why or why not?

   ____

   ____

   ____

   If this is possible, then construct the triangle below. Measure the angles and label them within your triangle.

4. With your partner, discuss the possibility of drawing a triangle in which you have more than one obtuse angle. Is this possible? __________ Explain why or why not.

   ____

   ____

   ____

(continued)

Name______________________________ Date ______________

ACTIVITY 13

Sums of Angles in a Triangle *(continued)*

5. In this step draw a horizontal straight angle through the vertex shown below. How many degrees are in your straight angle? __________

6. Now take your triangle from Step 1 and tear it into three pieces, making sure that you do not rip through any angle measure. Each piece must contain a colored corner. See the diagram below.

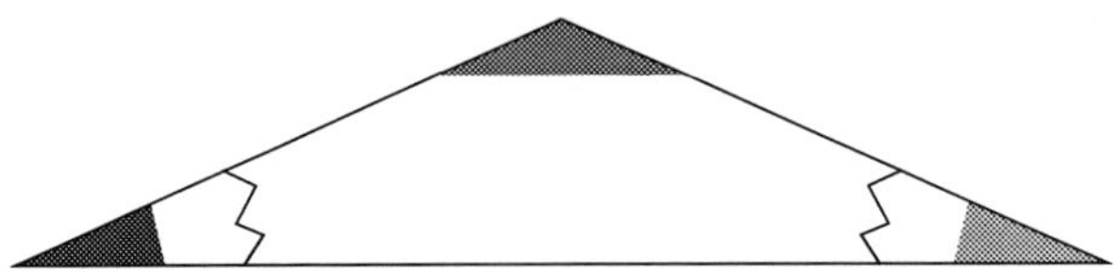

7. Using the straight angle that you drew in Step 5, align the angle pieces of your torn triangle on the straight angle, so that the angles abut each other and each angle vertex of the triangle is sharing the vertex of the straight angle. Tape the angles down using the vertex in Step 4. See the diagram below.

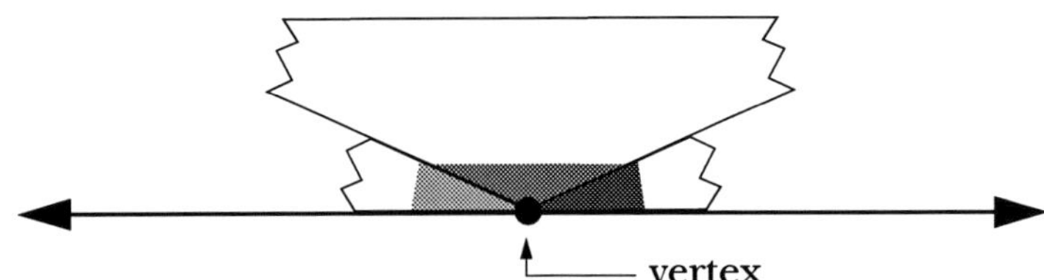

(continued)

Name________________________________ Date ________________

Activity 13

Sums of Angles in a Triangle *(continued)*

8. After you each have taped down your angle pieces, discuss with your partner what you observe about the angles. Write your discoveries below. Be prepared to share your findings with the class.

Triangles

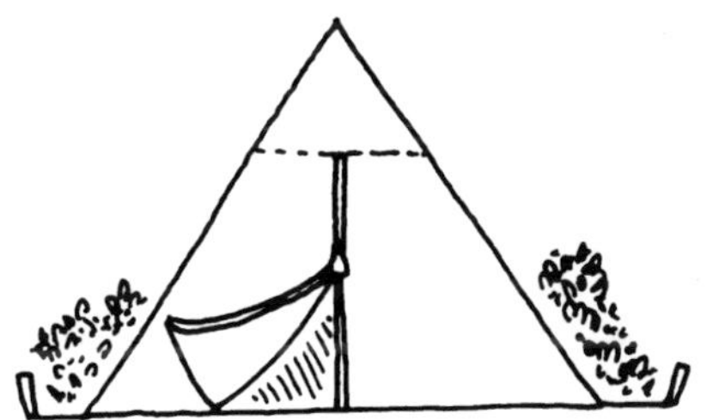

ACTIVITY 14

Similar Triangles

Teacher Page

Learning Outcomes

Students will be able to

- solve for unknown heights using proportions.
- measure various lengths with appropriate measuring tools.

Overview

Student groups will calculate the height of buildings, flagpoles, or trees by measuring the objects' shadows and solving a proportion formula. Students will also calculate each other's heights using the same method.

Time Requirements

30 minutes

Group Size

Pairs or groups of three

Materials

- rulers
- tape measures
- lab sheets
- calculators

Procedure

1. This lab works best outdoors during a sunny day. If you don't have these conditions, then you could use a lamp with a bright light to cast shadows.
2. Sometimes building shadows may be hard to measure. Students must measure the perpendicular distance of the shadow. If your site has trees or flagpoles, they may work better.
3. When students measure their heights, the calculated heights may not match their exact heights. Discuss why this may happen. Some students may question where to measure the shadow, from the back of the student's foot or in the front. Have students experiment to see which gives the more accurate measure. Make sure the students set up their proportions correctly and in the correct order.

Name______________________________________ Date ____________________

Activity 14

Similar Triangles

Part One

In this lab, we are going to try to figure out the height of a building or flagpole using the process of similar figures and proportions.

1. Go outside and measure the shadow of a building or flagpole in feet.

   Measure of shadow of building or flagpole: ____________________

2. Now place a ruler perpendicular to the ground and measure its shadow.

   Measure of shadow of ruler in inches: ____________________

3. Change the measure of the ruler's shadow to feet by dividing by 12.

   Measure of ruler's shadow in feet: ____________________

4. You are now ready to calculate the height of the building or flagpole.

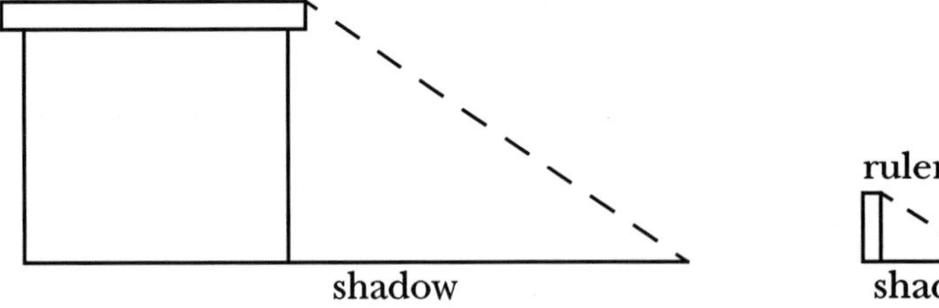

Proportion: $\dfrac{\text{Height of building}}{\text{Length of building's shadow}} = \dfrac{\text{Height of ruler}}{\text{Length of ruler's shadow}}$

Fill in with your measurements: $\dfrac{\text{Height of building \_\_\_\_\_\_}}{\text{Length of shadow \_\_\_\_\_\_}} = \dfrac{\text{Height of ruler \_\_\_\_\_\_}}{\text{Length of shadow \_\_\_\_\_\_}}$

What is your calculated building height? ____________________

Part Two

1. Have your partner measure the length of your shadow and try to calculate your height.

2. Use the proportion of the length of the ruler to its shadow as the other part of the proportion. Does your calculated height come close to your actual height? ________

ACTIVITY 15

Pythagorean Theorem

Teacher Page

Learning Outcome

Students will be able to apply the Pythagorean theorem to real-life situations.

Overview

Students will use a variety of hands-on methods to learn about, test, and calculate with the Pythagorean theorem.

Time Requirements

40–60 minutes

Group Size

Pairs

Materials

For each group

- 25 one-inch-square tiles
- 3" × 4" × 5" right triangles
- 26-inch piece of string
- centimeter rulers
- centimeter dot-matrix paper

Procedure

Part One

1. Pass out a 3" × 4" × 5" right triangle and 25 one-inch square tiles
2. Have groups build squares on each of the legs of the right triangle. If they do it correctly, all 25 tiles will be used, with 9 on the 3-inch side and 16 on the 4-inch side.

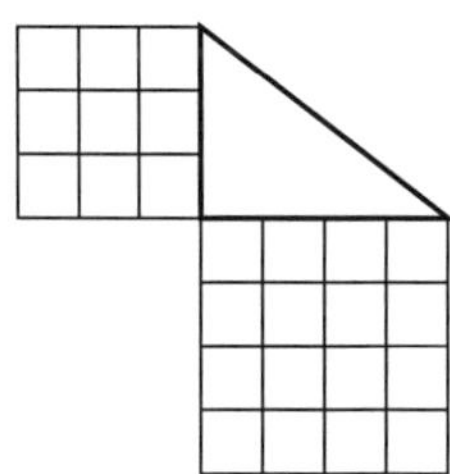

3. Have groups take the tiles and build a "square" on the hypotenuse using as many of the 25 tiles as they need. They will see that all the tiles will be used on the hypotenuse. This should show students that the sum of the squares of the legs equals the square formed on the hypotenuse.

Part Two

1. In this activity, the students are using the converse of the Pythagorean theorem. That is, if the squares of two sides of a triangle equal the square of the hypotenuse, then the triangle must be a right triangle.
2. Using the 6-8-10-inch string, the groups will locate right triangles in the room and use the string to prove the triangles are right triangles. To do this, students will put the

6-inch length on one leg of the triangle and the 8-inch length on the other leg. The remaining 10-inch length should form the hypotenuse.

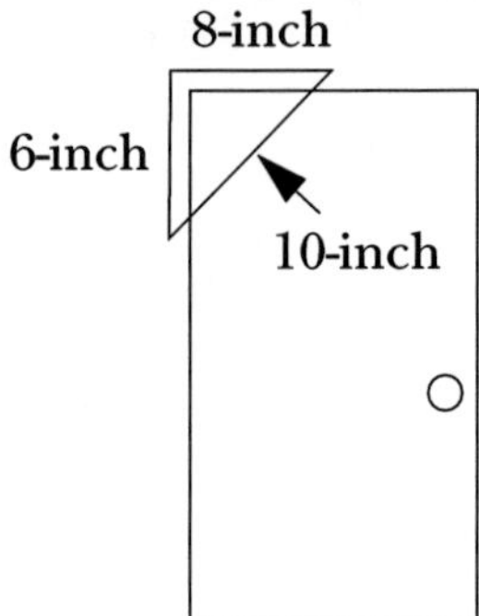

3. The groups should also try an object that doesn't have a right angle to show that the three pieces will not fit together.

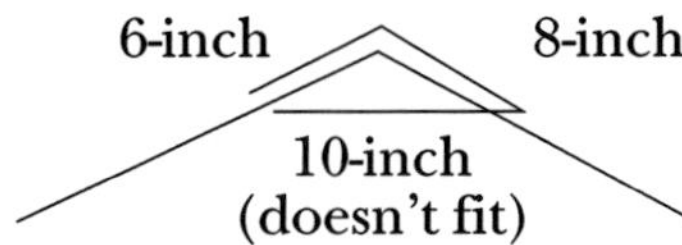

Part Three

1. Using dot-matrix paper students make several right triangles.
2. They then measure the legs and put the measurements into the Pythagorean formula to determine the length of the hypotenuse.
3. Teams then measure the hypotenuse with a ruler to verify the results. This activity gives students practice with the Pythagorean theorem with visual results to verify their work.

Part Four

In this activity, the student will begin to see a spiral form. The key is to keep one of the legs constant with a length of one. The other leg is the hypotenuse of the previous right triangle.

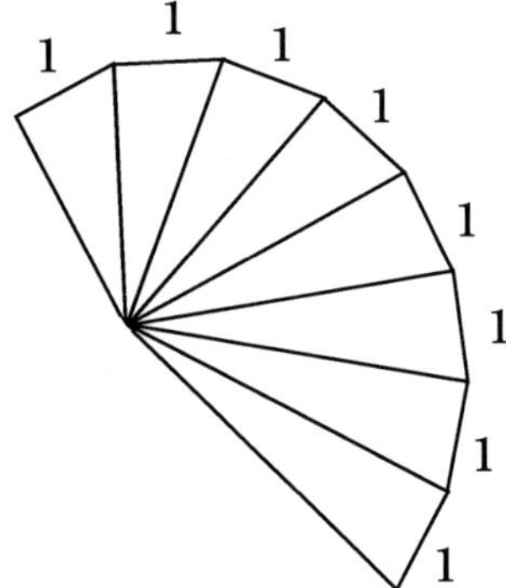

Name______________________________ Date ______________

ACTIVITY 15

Pythagorean Theorem

Part One

Your teacher will give you a picture of a right triangle and some 1-inch tiles. Count out 25 tiles. Determine how long the **"legs"** (the sides of the triangle that form a right angle) are in tiles.

One leg is __________ tiles long.

The other leg is __________ tiles long.

Now form squares on each of these legs with the tiles. ***Example:*** If the leg were 6 tiles long, the square would be 6 tiles by 6 tiles. Fill in this square with the remaining tiles. If you do this correctly, you will use all 25 tiles.

Now look at the "slanted side" of the triangle. The slanted side is called the ***hypotenuse.*** What do you *estimate* the length of the hypotenuse to be in tiles?

Estimated length is __________ tiles.

Now use your tiles to see how long the hypotenuse is.

The hypotenuse is __________ tiles long.

Following the same procedure as above, use as many tiles as you need to make a square on the hypotenuse and fill it in with the tiles.

What do you notice?

____

____

Can you make a rule about the legs and hypotenuse of a right triangle from what you have observed? Please state it below.

____

____

The **Pythagorean theorem** states that if a triangle is a right triangle, then:

$$a^2 + b^2 = c^2$$

Explain in words what this means:

____

____

(continued)

Name________________________________ Date ____________________

Activity 15

Pythagorean Theorem *(continued)*

Part Two

The **converse** of the Pythagorean theorem states that if a triangle has sides of length *a, b,* and *c,* and $a^2+b^2=c^2$, then the triangle is a right triangle. This means that you can determine whether a triangle has a right angle by testing the sides in the formula.

The converse is used in laying pipe, constructing houses, or anything that requires formation of a right angle.

Take a 26" piece of string, measure 6 inches, and place a knot at the 6" mark. From that knot, measure 8 inches and knot again. The remaining section should be 10", so cut off any leftover string.

Could 6, 8, and 10 be the measures of the sides of a right triangle? ____________

How do you know? ____

____

Using this string, find places around the room that appear to form right angles and use this measuring device to check them. List below the angles you examined, and whether they were right angles.

____

____

____

Now find an angle that is not a right angle, and test to see that the 6-8-10 string will not work.

(continued)

Name____ Date ____________________

ACTIVITY 15

Pythagorean Theorem *(continued)*

Part Three

On your matrix paper, draw a right triangle by connecting dots up and down or right and left. The legs can be any length you want.

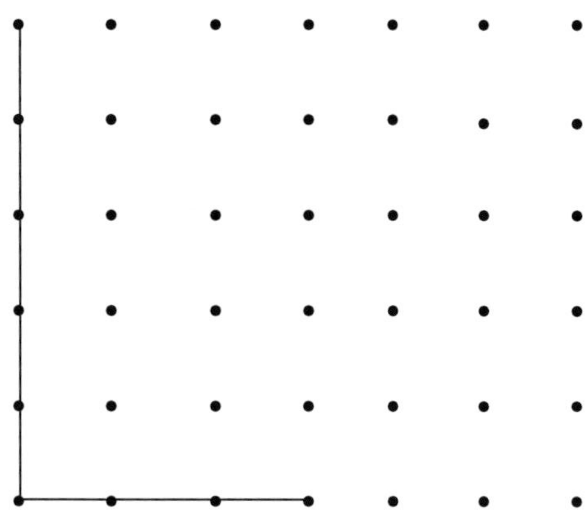

Then connect the diagonal (the hypotenuse).

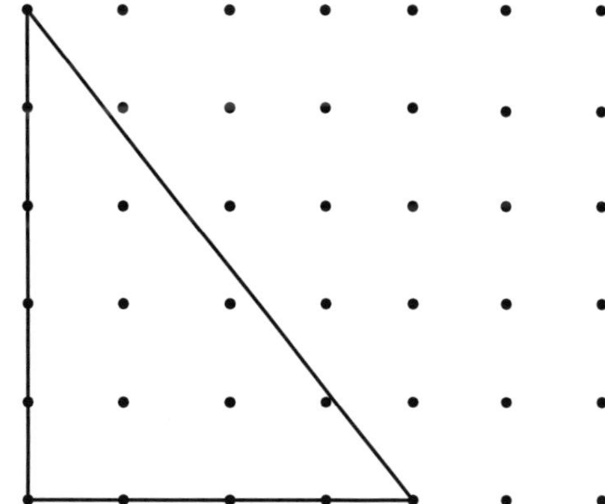

(continued)

Name____ Date ____________________

ACTIVITY 15

Pythagorean Theorem *(continued)*

Test the lengths with the Pythagorean theorem. In this example, the triangle drawn has the lengths of 4 and 5 cm. Plug these lengths into the Pythagorean formula to find the length of the hypotenuse.

$a^2 + b^2 = c^2$

as $4^2 + 5^2 = c^2$

$16 + 25 = c^2$

$41 = c^2$

$\sqrt{41} = \sqrt{c^2}$

$6.4 = c$

Now measure the hypotenuse with your centimeter ruler to verify this solution.

Try several examples on the dot-matrix paper to demonstrate the Pythagorean theorem.

Part Four

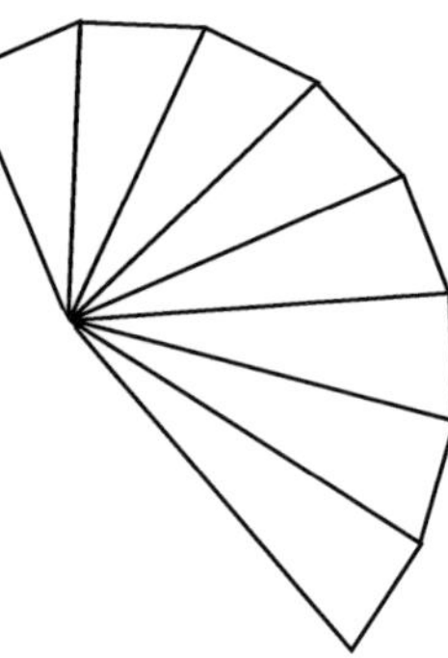

In nature, the spiral shell is formed by connected right triangles in which one leg stays one unit long, while the other leg increases to the length of the previous hypotenuse.

Look at the diagram on the right. Try creating this spiral on your grid paper.

See if you or any of your classmates can find a shell, or a photograph of a shell, that illustrates this.

ACTIVITY 16

Median Line of a Triangle

Teacher Page

Learning Outcomes

Students will be able to

- determine the midpoints of two sides of a triangle using a compass.
- discover that a segment whose endpoints are the midpoints of two sides of a triangle is parallel to the third side of the triangle, and its length is one half the length of the third side.

Overview

Through construction and measurement, students will discover properties of the line that connects the midpoints of two sides of a triangle.

Time Requirements

30–45 minutes

Group Size

Pairs

Materials

- compass
- protractor
- lab sheet
- ruler

Procedure

Before the lab, review the use of a compass with the class. Have them practice drawing circles first. Then introduce/review how you manipulate the compass to bisect line segments.

You may also need to review how to use a protractor to record angle measurements.

1. After reviewing how to bisect a line, have the students bisect line segments $\overline{MN}$ and $\overline{MP}$.
2. Next have the students connect the two midpoints found and identify these points as S and T, respectively.
3. When they measure $\angle MST$ and $\angle N$, they will find that they are equal.
4. The same will occur with $\angle P$ and $\angle MTS$. They will be equal.

5. Line segments $\overline{ST}$ and $\overline{NP}$ are parallel because if corresponding angles on the same side of a transversal are equal, the lines are parallel.
6. The students should notice that the measure of $\overline{ST}$ is half the measure of $\overline{NP}$.
7. Students will discover that this will hold true for any triangle regardless of the shape.

Name________________________________ Date ________________

ACTIVITY 16

Median Line of a Triangle

1. Use a compass to bisect line segments $\overline{MN}$ and $\overline{MP}$.

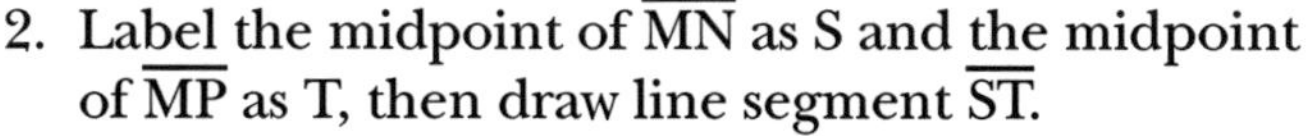

2. Label the midpoint of $\overline{MN}$ as S and the midpoint of $\overline{MP}$ as T, then draw line segment $\overline{ST}$.

3. Use a protractor and measure $\angle MST$ and $\angle N$. $\angle MST =$ ______________ $\angle N =$ ______________

4. Next measure $\angle P$ and $\angle MTS$. $\angle P =$ ______ $\angle MTS$ ________

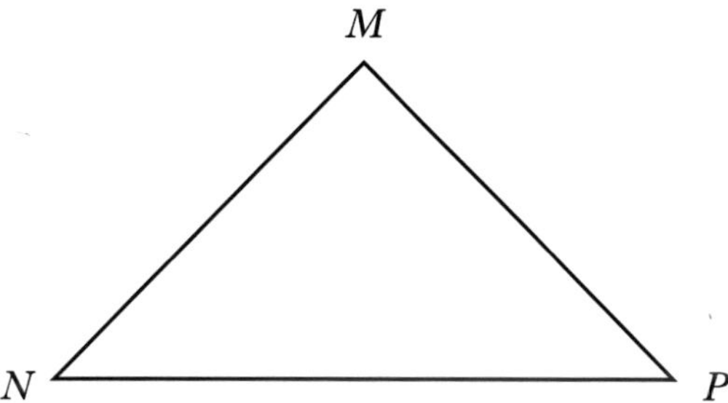

5. Compare the angle measures. What relationship does this suggest for line segments $\overline{ST}$ and $\overline{NP}$? ________________________________

____

____

Why? ____

____

6. Measure $\overline{ST}$ and $\overline{NP}$. How do these values compare?

____

____

____

(continued)

Name____ Date ______________________

ACTIVITY 16

Median Line of a Triangle *(continued)*

7. Repeat the above steps for the obtuse and right triangles below.

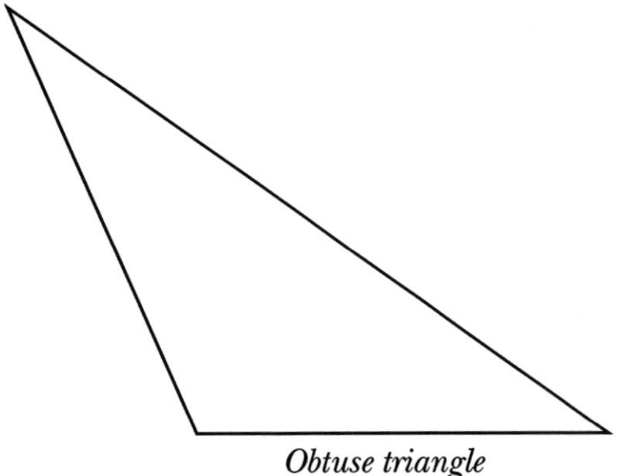

Obtuse triangle

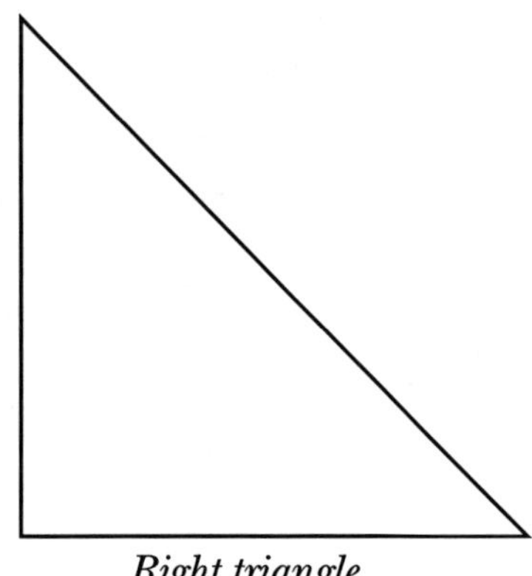

Right triangle

Do you get the same results? Explain.

____

____

____

____

Quadrilaterals

Activity 17

Investigating Special Quadrilaterals

Teacher Page

Learning Outcomes

Students will be able to

- measure side and angles with appropriate tools.
- discover special properties for regular quadrilaterals.

Overview

Students will measure the sides and angles of regular quadrilaterals and discover special characteristics for these quadrilaterals.

Time Requirements

30 minutes

Group Size

Pairs

Materials

- rulers
- protractors
- lab sheets

Procedure

Part One

1. Distribute lab sheets, rulers, and protractors to each team.
2. You may need to review how to read a protractor and measure in English units.
3. As you observe students' measurements, the following should occur:
   (a) The measure of the sides of the square and its angles should all be equal.
   (b) The opposite sides of the rectangle should be equal and each angle should measure 90 degrees.
   (c) The opposite sides of the parallelogram and the opposite angles of the parallelogram should be equal.
   (d) All sides of the rhombus should be equal.
   (e) The non-parallel sides of the isosceles trapezoid as well as its base angles should be equal.

Part Two

Possible Answers

1. The rhombus and the square have all sides equal.
2. Opposite sides of the parallelogram and rectangle are equal.
3. Opposite angles of the parallelogram and rectangle are equal.
4. One pair of opposite sides in the trapezoid and either pair of opposite sides in the rectangle or parallelogram are parallel.
5. All angles of the square and rectangle are 90-degree angles.
6. Consecutive angles of any regular quadrilateral are supplementary.

These discoveries lead to the following theorems. Since a rectangle, square, and rhombus are all parallelograms, the parallelogram theorems apply to these other shapes also.

(a) Opposite sides of a parallelogram are congruent.

(b) Opposite angles of a parallelogram are congruent.

(c) Consecutive angles of a parallelogram are supplementary.

(d) Both pairs of base angles of an isosceles trapezoid are congruent.

Name______________________________________ Date ____________________

ACTIVITY 17

Investigating Special Quadrilaterals

Part One

For the quadrilaterals below, measure the sides and the angles.

Square

A B

D C

Measure of side *AB* _______ Measure angle *A* _______

Measure of side *BC* _______ Measure angle *B* _______

Measure of side *CD* _______ Measure angle *C* _______

Measure of side *AD* _______ Measure angle *D* _______

Rectangle

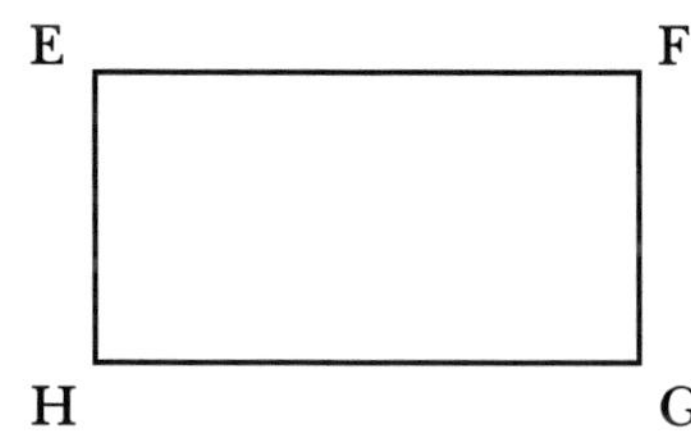

Measure of side *EF* _________ Measure angle *E* _________

Measure of side *FG* _________ Measure angle *F* _________

Measure of side *GH* _________ Measure angle *G* _________

Measure of side *EH* _________ Measure angle *H* _________

Parallelogram

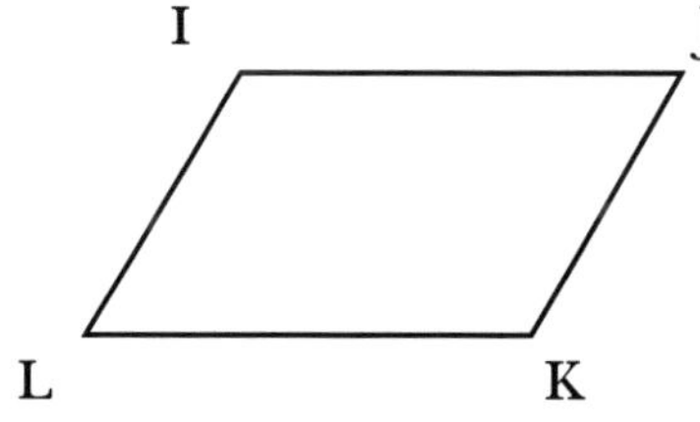

Measure of side *IJ* _______ Measure angle *I* _______

Measure of side *JK* _______ Measure angle *J* _______

Measure of side *KL* _______ Measure angle *K* _______

Measure of side *IL* _______ Measure angle *L* _______

(continued)

Name________________________________ Date ________________

ACTIVITY 17

Investigating Special Quadrilaterals *(continued)*

Rhombus

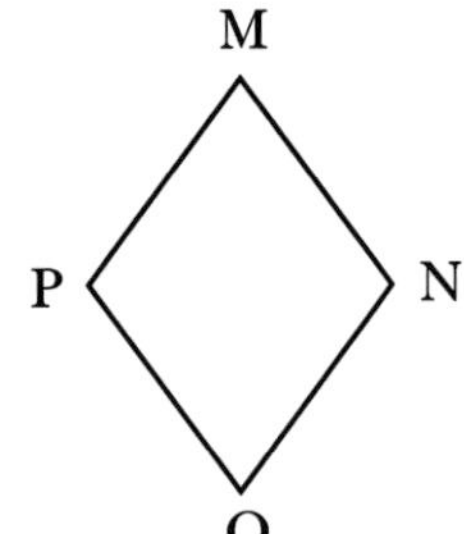

Measure of side *MN*_______ Measure angle *M*_______

Measure of side *NO*_______ Measure angle *N*_______

Measure of side *OP*_______ Measure angle *O*_______

Measure of side *PM*_______ Measure angle *P*_______

Trapezoid

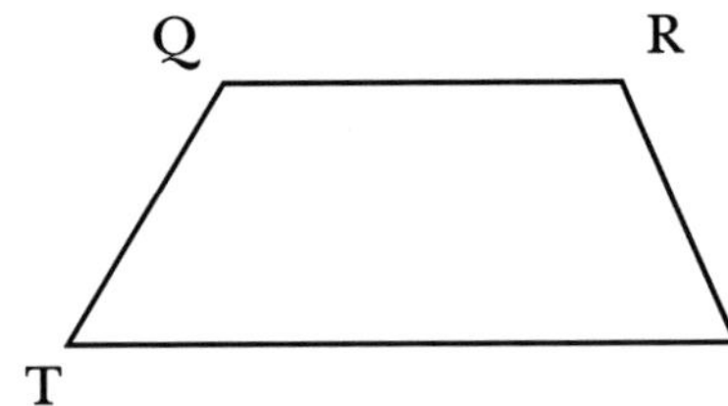

Measure of side *QR*_______ Measure angle *Q*_______

Measure of side *RS*_______ Measure angle *R*_______

Measure of side *ST*_______ Measure angle *S*_______

Measure of side *TQ*_______ Measure angle *T*_______

Part Two

List 5 characteristics that two or more of these special quadrilaterals share with each other.

1. ____
2. ____
3. ____
4. ____
5. ____

Activity 18

Diagonals and Special Quadrilaterals

Teacher Page

Learning Outcomes

Students will be able to

- measure line segments and angles with appropriate tools.
- discover special properties for diagonals of regular quadrilaterals.

Overview

Students will measure the diagonals, diagonal segments, and angles formed by diagonals and discover special properties concerning diagonals of regular quadrilaterals.

Time Requirements

45 minutes

Group Size

Pairs

Materials

- rulers
- protractors

Procedure

Part One

In this activity students will measure the line segments formed by the intersection of two diagonals of a parallelogram. They will discover that the diagonals of a parallelogram bisect each other.

Part Two

In this activity students will measure the diagonals for each of the rectangles in either centimeters or inches. They will discover that the diagonals of a rectangle are equal. They will also discover that the diagonals of the rectangle bisect each other.

Part Three

In this activity students will measure the angles formed by the intersection of the two diagonals in a rhombus. Students will discover that the angles formed are right angles and concur that the diagonals of a rhombus are perpendicular. They will also observe that the diagonals of a rhombus are equal and bisect each other.

In Summary

The diagonals of a parallelogram

- bisect each other.

The diagonals of a rectangle

- are equal.
- bisect each other.

The diagonals of a rhombus

- are perpendicular.
- are equal.
- bisect each other.

Name____ Date ____________________

ACTIVITY 18

Diagonals and Special Quadrilaterals

Part One

Measure the following line segments for the parallelograms below.

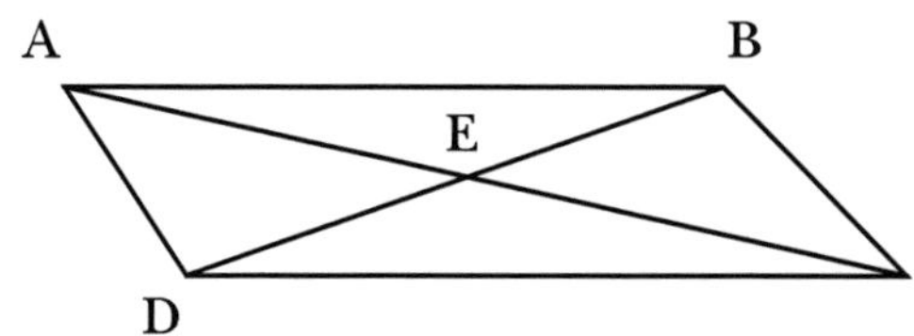

$\overline{AE}$ = ________ $\overline{EC}$ = ________

$\overline{DE}$ = ________ $\overline{EB}$ = ________

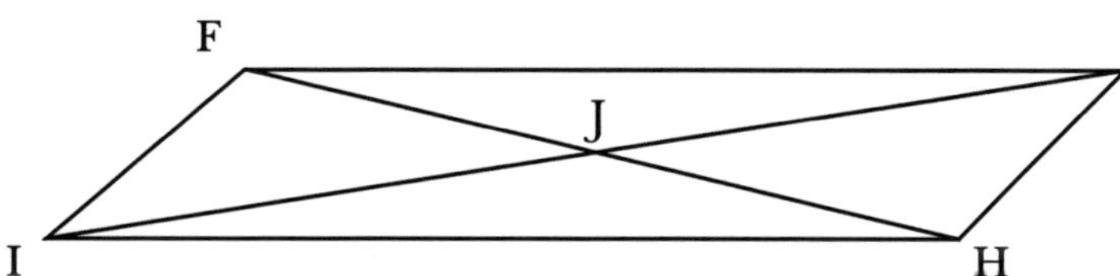

$\overline{FJ}$ = ________ $\overline{JH}$ = ________

$\overline{IJ}$ = ________ $\overline{JG}$ = ________

What observation can you make about diagonals of a parallelogram?

____

____

____

Part Two

Measure the diagonals for the following rectangles below.

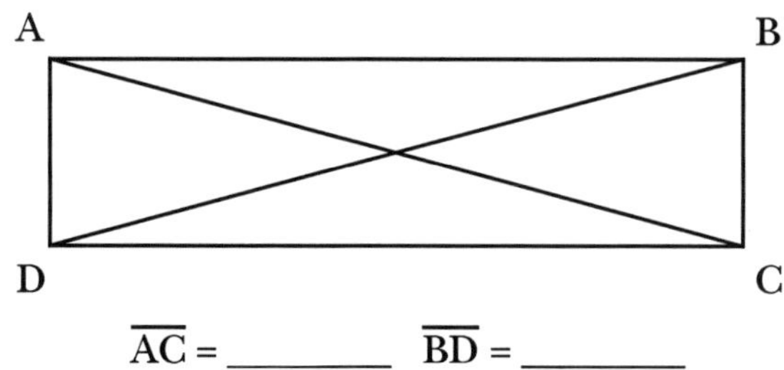

$\overline{AC}$ = ________ $\overline{BD}$ = ________

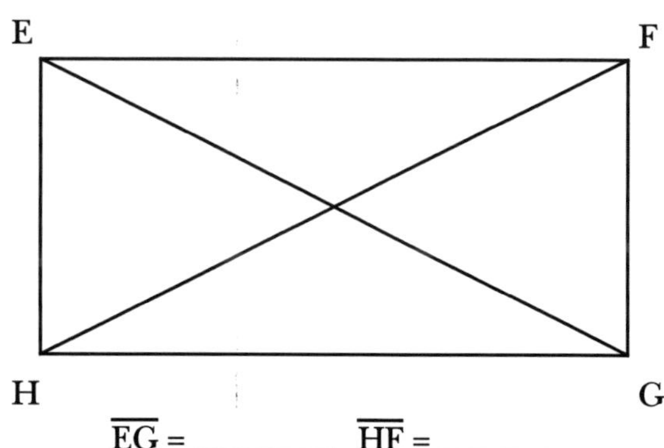

$\overline{EG}$ = ________ $\overline{HF}$ = ________

(continued)

Name____ Date ____________________

Activity 18

Diagonals and Special Quadrilaterals *(continued)*

What observation can you make about the diagonals of a rectangle?

____

____

Does the observation that you made about parallelograms in Part One hold true for the rectangles in Part Two?

Part Three

Measure the following designated parts of the rhombi below.

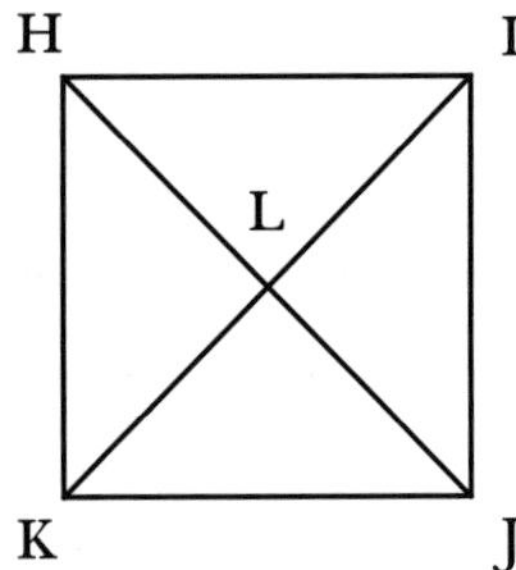

Measure the following angles:

$\angle HLK$ = ________

$\angle HLI$ = ________

$\angle ILJ$ = ________

$\angle JLK$ = ________

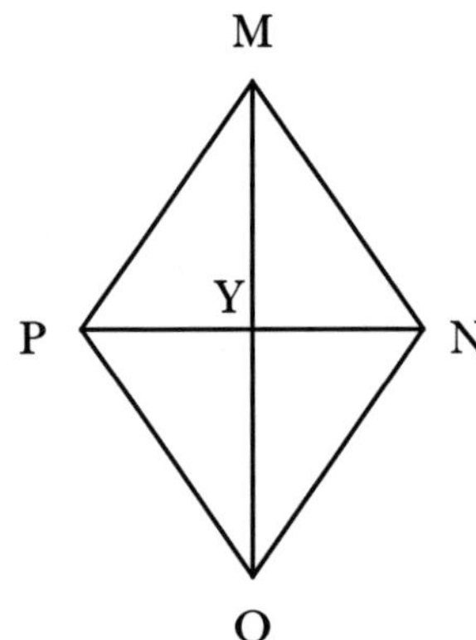

Measure the following angles:

$\angle PYM$ = ________ $\angle MYN$ = ________

$\angle PYO$ = ________ $\angle NYO$ = ________

What observation can you make about the diagonals of a rhombus?

____

____

Does the observation that you made about parallelograms in Part One hold true for the rhombi in Part Three?

Does the observation that you made about rectangles in Part Two hold true for the rhombi in Part Three?

(continued)

Name________________________________ Date ________________

ACTIVITY 18

Diagonals and Special Quadrilaterals *(continued)*

In Summary

Complete the following statements:

The diagonals of a parallelogram

- ______________________________

The diagonals of a rectangle

- ______________________________
- ______________________________

The diagonals of a rhombus

- ______________________________
- ______________________________
- ______________________________

Share Your Bright Ideas with Us!

We want to hear from you! Your valuable comments and suggestions will help us meet your current and future classroom needs.

Your name____Date____________________

School name________________________________Phone_________________________

School address____

Grade level taught________Subject area(s) taught______________________Average class size____

Where did you purchase this publication?____

Was your salesperson knowledgeable about this product? Yes_____ No_____

What monies were used to purchase this product?

___School supplemental budget ___Federal/state funding ___Personal

Please "grade" this Walch publication according to the following criteria:

Quality of service you received when purchasing	A	B	C	D	F
Ease of use	A	B	C	D	F
Quality of content	A	B	C	D	F
Page layout	A	B	C	D	F
Organization of material	A	B	C	D	F
Suitability for grade level	A	B	C	D	F
Instructional value	A	B	C	D	F

COMMENTS:____

What specific supplemental materials would help you meet your current—or future—instructional needs?

Have you used other Walch publications? If so, which ones?________________________________

May we use your comments in upcoming communications? ___Yes ___No

Please **FAX** this completed form to **207-772-3105**, or mail it to:

Product Development, J.Weston Walch, Publisher, P.O. Box 658, Portland, ME 04104-0658

We will send you a **FREE GIFT** as our way of thanking you for your feedback. **THANK YOU!**